International and Development Education

The *International and Development Education Series* focuses on the complementary areas of comparative, international, and development education. Books emphasize a number of topics ranging from key international education issues, trends, and reforms to examinations of national education systems, social theories, and development education initiatives. Local, national, regional, and global volumes (single authored and edited collections) constitute the breadth of the series and offer potential contributors a great deal of latitude based on interests and cutting edge research. The series is supported by a strong network of international scholars and development professionals who serve on the International and Development Education Advisory Board and participate in the selection and review process for manuscript development.

SERIES EDITORS
John N. Hawkins
Professor Emeritus, University of California, Los Angeles
Senior Consultant, IFE 2020 East West Center

W. James Jacob
Assistant Professor, University of Pittsburgh
Director, Institute for International Studies in Education

PRODUCTION EDITOR
Heejin Park
Project Associate, Institute for International Studies in Education

INTERNATIONAL EDITORIAL ADVISORY BOARD
Clementina Acedo, *UNESCO's International Bureau of Education, Switzerland*
Philip G. Altbach, *Boston University, USA*
Carlos E. Blanco, *Universidad Central de Venezuela*
Sheng Yao Cheng, *National Chung Cheng University, Taiwan*
Ruth Hayhoe, *University of Toronto, Canada*
Wanhua Ma, *Peking University, China*
Ka-Ho Mok, *University of Hong Kong, China*
Christine Musselin, *Sciences Po, France*
Yusuf K. Nsubuga, *Ministry of Education and Sports, Uganda*
Namgi Park, *Gwangju National University of Education, Republic of Korea*
Val D. Rust, *University of California, Los Angeles, USA*
Suparno, *State University of Malang, Indonesia*
John C. Weidman, *University of Pittsburgh, USA*
Husam Zaman, *Taibah University, Saudi Arabia*

Institute for International Studies in Education
School of Education, University of Pittsburgh
5714 Wesley W. Posvar Hall, Pittsburgh, PA 15260 USA

Center for International and Development Education
Graduate School of Education & Information Studies, University of California, Los Angeles
Box 951521, Moore Hall, Los Angeles, CA 90095 USA

Titles:
Higher Education in Asia/Pacific: Quality and the Public Good
Edited by Terance W. Bigalke and Deane E. Neubauer

Affirmative Action in China and the U.S.: A Dialogue on Inequality and Minority Education
Edited by Minglang Zhou and Ann Maxwell Hill

Critical Approaches to Comparative Education: Vertical Case Studies from Africa, Europe, the Middle East, and the Americas
Edited by Frances Vavrus and Lesley Bartlett

Curriculum Studies in South Africa: Intellectual Histories & Present Circumstances
Edited by William F. Pinar

Higher Education, Policy, and the Global Competition Phenomenon
Edited by Laura M. Portnoi, Val D. Rust, and Sylvia S. Bagley

The Search for New Governance of Higher Education in Asia
Edited by Ka-Ho Mok

International Students and Global Mobility in Higher Education: National Trends and New Directions
Edited by Rajika Bhandari and Peggy Blumenthal

Curriculum Studies in Brazil: Intellectual Histories, Present Circumstances
Edited by William F. Pinar

Access, Equity, and Capacity in Asia Pacific Higher Education
Edited by Deane Neubauer and Yoshiro Tanaka

Policy Debates in Comparative, International, and Development Education
Edited by John N. Hawkins and W. James Jacob

Increasing Effectiveness of the Community College Financial Model: A Global Perspective for the Global Economy
Edited by Stewart E. Sutin, Daniel Derrico, Rosalind Latiner Raby, and Edward J. Valeau

Curriculum Studies in Mexico: Intellectual Histories, Present Circumstances
William F. Pinar

Internationalization of East Asian Higher Education: Globalization's Impact
John D. Palmer

Taiwan Education at the Crossroad: When Globalization Meets Localization
Chuing Prudence Chou and Gregory S. Ching

Mobility and Migration in Asian Pacific Higher Education
Edited by Deane E. Neubauer and Kazuo Kuroda

University Governance and Reform: Policy, Fads, and Experience in International Perspective
Edited by Hans G. Schuetze, William Bruneau and Garnet Grosjean

Higher Education Regionalization in Asia Pacific: Implications for Governance, Citizenship and University Transformation
Edited by John N. Hawkins, Ka-Ho Mok and Deane E. Neubauer

Post-Secondary Education and Technology: A Global Perspective on Opportunities and Obstacles to Development
Edited by Rebecca A. Clothey, Stacy Austin-Li, and John C. Weidman

Education and Global Cultural Dialogue: A Tribute to Ruth Heyhoe
Edited by Karen Mundy and Qiang Zha

ALSO BY AUTHORS

Frances Vavrus
Desire and Decline: Schooling amid Crisis in Tanzania (2003)

Lesley Bartlett
The Word and the World: The Cultural Politics of Literacy in Brazil (2009)

CRITICAL APPROACHES TO COMPARATIVE EDUCATION

VERTICAL CASE STUDIES FROM AFRICA, EUROPE, THE MIDDLE EAST, AND THE AMERICAS

EDITED BY
FRANCES VAVRUS AND LESLEY BARTLETT

CRITICAL APPROACHES TO COMPARATIVE EDUCATION

Copyright © Frances Vavrus and Lesley Bartlett, 2009.

All rights reserved.

First published in hardcover in 2009 by PALGRAVE MACMILLAN® in the United States—a division of St. Martin's Press LLC, 175 Fifth Avenue, New York, NY 10010.

Where this book is distributed in the UK, Europe and the rest of the world, this is by Palgrave Macmillan, a division of Macmillan Publishers Limited, registered in England, company number 785998, of Houndmills, Basingstoke, Hampshire RG21 6XS.

Palgrave Macmillan is the global academic imprint of the above companies and has companies and representatives throughout the world.

Palgrave® and Macmillan® are registered trademarks in the United States, the United Kingdom, Europe and other countries.

ISBN: 978–1–137–36654–2

Library of Congress Cataloging-in-Publication Data is available from the Library of Congress.

A catalogue record of the book is available from the British Library.

Design by Newgen Knowledge Works (P) Ltd., Chennai, India.

First PALGRAVE MACMILLAN paperback edition: November 2013

10 9 8 7 6 5 4 3 2 1

To our colleagues at Teachers College, past and present,
whose research inspired our own and whose support made
this volume possible

Contents

Foreword

Henry M. Levin

Most publications in comparative education address either micro-units or macro-units of education, individual families, schools, and communities, or the overall system of education. Case studies are typically used to incorporate rich descriptions of a single school or community. However, attempts to characterize an entire system of education focus little on the details of individual schools and what happens inside them. The connections between these two polar perspectives are rarely analyzed in any detail. Yet, in an age of decentralization of resources and decision making and devolution of authority, the connections between the large and the small are crucial determinants of educational dynamics and school functioning. How change occurs in one part of the educational system can affect other parts as well.

The duality implicit in traditional approaches to comparative education creates the impressions that the micro and macro versions are only loosely connected. This was an insight that struck me when I arrived as a faculty member in 1968 at Stanford University. Although I was trained in a parochial version of economics, I had a curiosity about what went on in other disciplines in education. The large crowd of students queuing to enter Cubberley Hall on Mondays and Wednesdays at 11:00 a.m. induced me to enter one day, far too late to find a seat. In an auditorium with a capacity of 500, at least 600 students were jammed into seats, aisles, floors, stairways, and the stage, listening to a middle-aged couple talk about understanding education in the context of something called cultural transmission. My colleagues, George and Louise Spindler, stood before this adoring crowd displaying photographs, film strips, and artifacts of many societies demonstrating how education must be understood in its function of preparing the young for roles specific to their cultures.

Although I was time-deprived as an untenured faculty member under heavy pressure to publish and teach (in addition to family demands with three children and anti-Vietnam War activity), I attended as many of the Spindlers' classes as I could. In each, I learned about the detailed scrutiny of anthropologists and how they construct an understanding of culture by examining the interactions between powerful beliefs and the varied functions of institutions.

In contrast with this approach dwelling on the local and micro-aspects of education, the comparative education courses we taught from the perspective of economics and politics were very macro. In those courses, we looked at the overall system of education and how it was organized and financed to discharge its responsibilities. The details of how schools and communities functioned were rarely described in our comparative analyses other than some of their organizational features. Furthermore, as much of the literature on comparative education was bifurcated between the large and the small, so were the courses that we taught.

A major uniqueness of this volume is its attempt to connect these levels by constructing "vertical case studies," a term that the editors and authors use to address the flow of action across levels as influenced by political, social, economic, and cultural forces. One issue addressed through vertical case studies is the influence of neoliberalism as it reverberates through every level of the educational system. Of particular focus is the democratization of educational politics and practices as they are transformed through privatization, decentralization, and changes in participation. Changes in educational practices and organization are connected in these chapters to immigration policies, the War on Terror, the racialization of minority youth, and other socioeconomic, gender, and geographical hierarchies that are interconnected in their educational treatments. Further, the volume examines the influence of international development organizations on national and local educational practices, and the ways that international policies are appropriated and adopted.

As a Teachers College faculty member, I am proud that the individual contributors are former students and colleagues. Many of our faculty and students in the Department of International and Transcultural Studies have attempted to employ a critical approach to the issues addressed here. The leadership of the editors, Frances Vavrus and Lesley Bartlett, in their own research and teaching has inspired their colleagues at this institution and beyond it to consider far richer and broader ways to address their educational interests. In so many important respects, this volume represents the culmination of a decade of developments in comparative education that I hope will be extended within the field. As the world faces a fiscal crisis that is likely to alter considerably the face of education in coming decades, such vertical approaches will become increasingly important to understanding educational policy, programs, and practices. It is my expectation that the breakthroughs made in developing and applying vertical case studies to the wide range of issues in this book will be powerful tools for understanding other issues and other settings, thus making important contributions to comparative and development education.

Series Editors' Introduction

John N. Hawkins and
W. James Jacob

It is a pleasure to add to the *International & Development Education* series the new title: *Critical Approaches to Comparative Education*, edited by Frances Vavrus and Lesley Bartlett. We believe this volume will join those classics that have preceded it in helping to define and guide the field of Comparative and International Education. Together with a distinguished group of scholars who bridge theory and practice, they have produced a volume that will deepen our understanding of the field itself as well as the specific topics addressed in each chapter. Specifically, the introduction makes a cogent argument about the kinds of contributions qualitative research can make to the field of comparative education, and the subsequent chapters exemplify the value of such an approach. The *qualitative* focus of the project and the elaboration on the methodology of vertical and horizontal case studies is a welcome addition to the literature in the field of comparative and international education. A rethinking and redefinition of some critical concepts in the field is a hallmark of the volume and will serve to provoke further discussion of the direction of the field. The methodological issues raised in the book go beyond the more simplistic dichotomies of qualitative versus quantitative, or disciplinary versus inter-disciplinary (including social science silos versus area studies) and instead broaden our vision by weaving the complexity of globalization with more grounded vertical and horizontal case studies. Conceptually, the chapters demonstrate the powerful contributions to the field made by contemporary theoretical approaches, including actor-network theory, sociocultural analyses of policy, and the sociocultural concept of policy and practice as *bricolage*. We are certain that this volume will provoke a broad discussion of how we "know" and specifically, as the editors state in their introduction, how we know comparatively.

Acknowledgment

We would like to acknowledge the contributions of Casey Stafford and Jamie Weiss to this volume.

Abbreviations

ABC	Abstinence, Be Faithful, Use Condom
AGED	*Association Générale des Etudiants de Dakar* [General Association of Students of Dakar]
ALP	Accelerated Learning Program
BHC	Blossom Hill College
BIE	Bilingual Intercultural Education
CDA	Critical Discourse Analysis
CEO	County Education Officer
CES	Coalition of Essential Schools
CFV	Critical Friend Visit
CIE	Comparative and International Education
CIV	Community Information Volunteer
CNES	*Concertation Nationale sur l'Enseignement Supérieur* [National Consultation for Higher Education]
CPI	Corruption Perception Index
CSPD	Child Survival Protection and Development
DED	District Executive Director
DEO	District Education Officer
DES	Department of Education and Science
DOE	Department of Education
EFA	Education for All
EJA	*Escola para Jovens e Adultos* [School for Young People and Adults]
ESEA	Elementary and Secondary Education Act
ESRI	Economic and Social Council
EU	European Union
FNLA	National Front for the Liberation of Angola
GTZ	*Gesellschaft für Technische Zusammenarbeit* [German Technical Cooperation]
HIP	Health Information Project
HIV/AIDS	Human Immunodeficiency Virus/Acquired Immune Deficiency Syndrome

HPA	Humanities Preparatory Academy
IDP	Internally Displaced Person
IEA	International Association for the Evaluation of Educational Achievement
IHED	*Institut des Hautes Etudes de Dakar* [Institute of Higher Education of Dakar]
IMF	International Monetary Fund
INEE	International Network for Education in Emergencies
INGO	International Nongovernmental Organization
INS	Immigration and Naturalization Services
JBS	James Baldwin School
LGA	Local Government Authority
LGRP	Local Government Reform Program
MOE	Ministry of Education
MOU	Memorandum of Understanding
MPLA	Popular Liberation Movement of Angola
NCCA	National Council for Curriculum and Assessment
NCLB	No Child Left Behind
NGO	Nongovernmental Organization
NPAR	National Action Plan against Racism
NRC	Norwegian Refugee Council
NSEERS	National Entry-Exit Registration System
OMI	Office of the Minister of Integration
PAES	*Projet d'Amélioration de l'Enseignement Supérieur* [Higher Education Improvement Project]
PAID	Poverty Alleviation Initiative Developer
PBAT	Project Based Assessment Task
PEDP	Primary Education Development Programme
PEER	Programme for Education in Emergencies and Reconstruction (UNESCO)
PISA	Programme for International Student Assessment
PRSP	Poverty Reduction Strategy Paper
PT	*Partido dos Trabalhadores* [Workers' Party]
SES	Supplemental Educational Service
SIDA	Swedish International Development Cooperation Agency
SINI	Schools in Need of Improvement
SMED	*Secretario Municipal de Educaçao* [Municipal Secretary of Education]
SRH	Sexual and Reproductive Health
SSP	Small Schools Project
STI	Sexually Transmitted Infection
TEP	Teacher Emergency Package

UCAD	*Université Cheikh Anta Diop* [Cheikh Anta Diop University]
UN	United Nations
UNESCO	United Nations Educational, Scientific and Cultural Organization
UNICEF	United Nations Children's Fund
UNITA	National Union for the Total Independence of Angola
URT	United Republic of Tanzania
USAID	United States Agency for International Development

Introduction

Knowing, Comparatively

Lesley Bartlett and Frances Vavrus

The field of Comparative and International Education (CIE) has long had an uneasy relationship with one of its central concepts: comparison. Scholars would likely agree that a study involving more than two countries is comparative, but what about multisited case studies in a single country? While such studies of education in Angola or Lebanon would qualify as international by North Atlantic standards, are they also comparative? Some would argue they are not. For example, in his presidential address to the Comparative and International Education Society, Carnoy (2006) asserted, "[A]lthough individual country case studies can be implicitly comparative, the best comparative research compares similar interventions, outcomes, processes, and issues across countries and uses similar methodology and data collection" (554). In contrast, others argue that qualitative research in international education entails inherent comparison because the norms from one's own country cannot help but influence how another system is understood (see, e.g., Gingrich 2002). Further, some scholars have critiqued the ways in which rigid conceptualizations of comparison have regulated the production and uses of educational knowledge. For example, responding to Carnoy's statement earlier, Levin (2006, 576) wrote,

> Comparative studies must not be a straitjacket for describing, explaining, and evaluating educational phenomena in different settings. Diversity cannot be extruded into similar methods, measures, and comparative interpretations. Does using PISA [Programme for International Student Assessment] data really help us understand much about educational development in Chiapas in the south of Mexico with its rural, impoverished, largely indigenous population and Nuevo Leon in the north with its urban, relatively prosperous, industrial population and U.S. orientation? Should we restrict comparative analysis to the limited dimensions dictated by government, NGO, and multinationals with their own narrow agendas and interests?[1]

Further, as a field, CIE is not quite sure how to define its other key demarcation: international. Studies conducted in "another country" normally qualify, but this feeble definition soon falters because it rests on the speaker's point of view. If one lives in England, studies of American schooling are international; if one lives in the United States, they generally are not. An American scholar sees research in Brazil as fully international, but what is the Brazilian Comparative Education Society to do? Further, do studies of immigrant populations living in Ireland or the United States meet the criteria for inclusion in the field? What about studies of pedagogy or policy originating elsewhere—Freirean pedagogy from Brazil, for example—used in one's home country?

Finally, the field of education, at least in the United States, maintains a certain ambivalence about the concept of comparison while generally avoiding the international. Although rankings of the country's students on international math and reading examinations often make headlines, schools of education have long focused on domestic matters. When they have programs in comparative and international education, academic institutions tend to segregate faculty and students who study educational issues beyond the nation's borders from those who study similar phenomena at home, as though U.S. teachers and students are divorced from the international.

Instead of taking this national-international distinction at face value, *Critical Approaches to Comparative Education: Vertical Case Studies from Africa, Europe, the Middle East, and the Americas* posits that comparison should be central to the study of education in the United States and in other countries. Moreover, we contend that qualitative case studies that compare actors, institutions, and policies as they circulate "vertically" and "horizontally" ought to be considered as central to CIE as multicountry studies. Using a series of what we term *vertical case studies* from Africa, Europe, the Middle East, and the Americas (including the United States), this book explores how educational policy, programming, and practice are shaped by and in turn influence local, national, and international forces. It brings together the work of a dynamic group of scholars whose research traces the flows of people, actions, ideas, texts, and discourses that shape educational policy and practice through schools, communities, city or district educational offices, ministries of education, and international development organizations. In the course of this work, the authors included here reconceptualize the comparative process and redraw the conventional boundaries of the field of education.

In this introduction, we seek to move from the more common terrain of methodology to the less familiar landscape of epistemology before again returning to more pragmatic questions of comparative methods. We believe that students and scholars of comparative and international education need to pay greater attention to epistemological issues related to *what* can be known about the world and *how* it can be known through comparative research before attending to the rules and procedures—the methods—used to gain such knowledge. As Masemann argued in her 1990 CIES presidential address, "Our conceptions of ways of knowing have limited and restricted the very definition of comparative education that we have taught to students and used in our own research and, indeed, have promulgated to practitioners" (465).

In what follows, we discuss the history of approaches to comparison in the field of CIE and make an epistemological case for the value of qualitative approaches to comparison. We then argue for a specific kind of qualitative research—vertical case studies. Drawing on contemporary theoretical approaches in anthropology and sociology, we outline the specific features of such an approach. The chapters that follow the introduction demonstrate the value of developing ethnographically informed case studies that compare across time and space.

Comparing Versions of Comparison in the Field

The inauguration of comparative education as an area of study in the late nineteenth century enshrined the nation-state as the central unit of analysis. Scholars focused primarily on the school-society relationship, with little attention to transactions within schools or other educational arenas. As Sadler wrote in 1900, "[T]he things outside the schools matter even more than the things inside the schools, and govern and interpret the things inside" (49). Early comparative historians, such as Sadler (1900), explored the relationships between school and society in European countries. Hans (1949) and Kandel (1933) employed cross-national methods to consider the influence of religious and political systems on education, respectively.

During the 1960s, the process of comparison changed in important ways. Comparison increasingly meant the study of schooling and societies in two or more nation-states, while the ascendancy of quantitative methods marginalized historical and cultural considerations in mainstream comparative education research. For instance, the venerable Bereday, who succeeded Kandel at Teachers College, Columbia University, elaborated the four steps for rigorous comparative studies in his influential 1964 text, *Comparative Methods in Education*: "First description, the systematic collection of pedagogical information in one country, then interpretation, the analysis in terms of social sciences, then juxtaposition, a simultaneous review of several systems to determine the framework in which to compare them, and finally comparison, first of select problems and then of the total relevance of education in several countries" (27–28). Though he encouraged the field to identify "laws" of education and social development, he also urged scholars to learn the languages and social contexts of the countries being studied (Hayhoe and Mundy 2008, 9). Bereday's students, Noah and Eckstein, famously declared that the field should move "toward a science of comparative education" (1969). They described their approach as follows:

> [We promoted] an effort to instruct students about empirical research and to turn away from the descriptive and often normative approach that characterized most comparative education courses at the time. In particular, we tried to show how it was both possible and enlightening to use comparative data to *test hypotheses* about the relation between education and social phenomena. We asked our students to assemble evidence sufficient to test the cross-national validity of...statements. (1998, 9)

During this period, which saw the emergence of the International Association for the Evaluation of Educational Achievement (IEA) studies and the merger of the fields of CIE, many researchers turned their attention to development education. They were concerned with technical questions related primarily to education and national development, specifically the "modernization" of the countries in the global South. One could say that, for many scholars at this time, the "central problem of comparison was believed to be technical and not *theoretical*" (Popkewitz and Pereyra 1993, 7).

Comparing Ethnographic Approaches in the Field

In the 1970s, the emergence of a variety of ethnographic approaches challenged the reigning positivist paradigm. The earliest reverberations in CIE were the result of research in anthropology and education completed by the Spindlers and their colleagues. This group of scholars was concerned with how cultures sustain themselves through education, broadly conceived to extend well beyond schooling and to include enculturation and socialization (Spindler 1959, 1963, 1974; see also Spindler 2000). Early work in educational ethnography focused on two themes: comparative studies of socialization in societies without mass schooling, and investigations of the impact of modern mass schooling on traditional models of cultural transmission. Masemann's 1976 publication, "Anthropological Approaches to Comparative Education," marked the beginning of this shift within CIE from the macrolevel to the microlevel, and from less to more attention to theoretical problems of comparison. Masemann argued that anthropological research held promise for the comparative study of education, especially studies that compared processes of socialization, schooling, and educational institutions in diverse cultural settings.

A critical review by Foley, also published in the mid-1970s, provided an excellent picture of nascent anthropological contributions to CIE. Foley highlighted anthropology's epistemological contribution to CIE in the form of:

> a hermeneutic tradition of analysis... [which is a] critical, reflexive way of knowing.... The natural science mode of inquiry makes numerous assumptions about "reality" and the investigator's relation to the "out-there reality." Such investigations are guided by further assumptions about causal relationships, objectivity, research technology and techniques, and the primacy of the scientific world view.... [In contrast,] to know in the ethnographic sense has meant learning situationally-based linguistic and role performances well enough to "survive" (culturally).... Personally replicating appropriate language and behavior in another culture is a much more demanding form of replicability than the split-half coefficient of a survey questionnaire.... Rarely is the construct, context, and predictive validity of a "model" tested with such empirical rigor as in the ethnographic mode of knowing. Good anthropological inquiry has always been experiential and reflexive and not merely technical. (1977, 313–314)

Foley lauded "philosophical and politically critical social science focused on human subjectivity as a creative, historical force" (313). Anthropological studies of education during this period focused on a variety of topics, including "the various ways that schooling reinforces ethnic, linguistic, and class inequalities," "the relatively important role that schools play in mediating structural and particularly acculturation effects on students," and "the role of school systems in cultural evolution" (317–319).

From the 1970s onward, four developments particularly influenced the conduct of ethnographic work in CIE. The first was the growing popularity of interpretivist and phenomenological approaches inspired by the work of cultural anthropologists such as Geertz. Rooted in a humanistic tradition, Geertz insisted that the ethnographer's job was to use "thick description" of local contexts to interpret how an informant experienced and understood patterns of meaning as embodied in symbols (1993, 5, 6, 14). Geertz was skeptical about attempts to generalize beyond specific contexts; he argued that "the essential task of theory building here is not to codify abstract regularities but to make thick description possible, not to generalize across cases but to generalize within them" (1993, 26). Geertz's interpretive insights reverberated throughout educational research in the 1970s and 1980s (see, e.g., Stenhouse 1979; Spindler and Spindler 1987). Enticed by the focus on meaning and symbols, as well as the anthropological tropes of holism and cultural relativism, this work reminded CIE scholars that social actions are stimulated by all sorts of motivations, not all of which might be considered predictable, reliable, or even rational.

The second major ethnographic shift in the field was occasioned by the paradigmatic revolution introduced by ethnomethodologists, who rejected positivist models of social life. Building on the sociology of knowledge (see Berger and Luckman 1967), which argued that knowledge is the historically rooted product of a social process of negotiation, interpretation, and representation, ethnomethodologists criticized the field of CIE for developing proxy measures of social life to the detriment of attention to the microlevel. Making an important epistemological point, Heyman felt that scholars should ask themselves "what is the nature of the phenomenon to be studied and what are the implications of that nature for both the way it can be studied and what shall count as knowledge about it" (1979, 243). He complained that too many comparative studies

> ignore what goes on in day-to-day interaction in schools. In normal social science fashion data are gathered on indicators for phenomena, rather than observing the phenomena themselves.... Comparative studies in education generally have not studied those factors in educational processes which can be directly observed, but have settled for the measurement of indicators for supposed social realities through the study of official statistical data, questionnaire responses, recruitment patterns, interest groups and so on.... [R]esearch based on the measurement of indicators used to stand for concepts which are then causally related using the formal deductive logical properties of constancy and identity are a gross distortion of the very social reality which comparativists seek to understand and reveal. (1979, 241–242)

Heyman recommended that scholars conduct "systematic observation and analysis of the microcosmic world of everyday life," which he thought could be achieved via audio- and videotaped interactions (1979, 245).

Anthropological case studies from this period reminded the field of CIE of the value of examining the politics of interaction (including research interactions) and the interpretative imperative of all social actors, including teachers and students. Unfortunately, as Foley (1977) noted, studies from this period too often failed to compare their work or to look beyond the microlevel. They abandoned the "effort to relate their findings to other field studies of schools" and failed to "study up" (meaning studying those with power) in either a theoretical or a descriptive sense (321). Few studies reported reliably about the overall formal school system, its organizational structure, and national sociopolitical context, and thus they neglected to generate data for detailed comparisons between formal curriculum, teacher certification, pedagogical methods, and a host of questions studied in educational foundations. Thus, phenomenologically and/or ethnomethodologically inspired qualitative studies provided important corrections to the work in the early period of anthropology of education, but they often lost sight of the value of cross-cultural comparison. As Spindler (2000) acknowledged, too often "school ethnography has been micro-analytic rather than holistic and it has been confined to our own schools without comparative reference, [which] can give us perspectives on our own schools and our assumptions about education" (181).

The third and, in our opinion, most provocative approach introduced to the field of CIE during this period was critical ethnography. Informed by Marxist and feminist theories of schooling and social change, these scholars promoted ethnography as an appropriate comparative education methodology and critical theory as a way to conceptualize uneven local and global distributions of power. Critical ethnography was heavily influenced not only by the new sociology of education but specifically by Bourdieu's work on social reproduction and forms of capital (see, e.g., Bourdieu and Passeron 1977; Bourdieu 1986) and Willis's (1981) ingenious ethnography explaining "how working class kids get working class jobs." Masemann (1982), for one, called for critical ethnography because it "investigate[s] the lived life of school without necessarily limiting the analysis to the actors perceptions' of their situations" (13). Such research situated the careful study of schooling in specific contexts within a broader sociopolitical context, thereby addressing the call to "study up" to the national or international levels. Inspired by Willis, a spate of "resistance and accommodation studies" emerged that exemplified the strengths and shortcomings of this approach (Marcus 1998, 42). Too often critical ethnography in CIE suffered from what Marcus (1998) called a "macro-micro world narrative structure" in which the researcher examined how monolithic, dehistoricized, and oversimplified systems such as capitalism shaped local social action, even as they were resisted (43). The perspectives of school-based actors became mere illustrations of these larger processes, and their actions metonyms for the economic system itself.

Although both theory and method gained prominence in the field of education as a result of critical studies of schooling, ethnographic research rarely provided an opportunity for explicit comparison because of its insistence on particular local contexts. Critical ethnographers were aware of how larger forces shaped local interactions, but they failed to subject those forces to the same careful scrutiny they focused on spatially delimited locales (Marcus 1989). In other words, these studies too often took historical, cultural, and social forces as explanations instead of as contexts. The chapters that follow in this volume demonstrate a critical ethnographic concern with political-economic analysis while being careful to specify how global forces and inter/national institutions affect social relations at the local level. Moreover, as a collection, they allow for the comparison of similar actors, discourses, institutions, and policies across national boundaries.

A fourth important development in ethnographic research in CIE concerns the impact of poststructural theory. Much has been written about this trend, including the politics of knowledge and the dilemmas of representation.[2] However, while various scholars have touted the conceptual and methodological benefits poststructural theory might offer CIE,[3] postmodern ethnography *per se* never edged its way into the field (Marginson and Mollis 2001). Nevertheless, the impact of Foucault and feminist poststructuralists is certainly felt in several of the chapters in this volume, whose authors interrogate specific power-knowledge couplings, theories of subjectivity, and discourses that systematically construct the subjects who use them and the social worlds of which they speak.

Comparing Vertically and Horizontally

In this volume, we build upon and extend the heritage of ethnographic approaches in the field of CIE, though our work is noticeably chastened by the dilemmas of poststructuralism. Informed by phenomenologists, ethnomethodologists, and critical ethnographers, we argue that contextual knowledge, including attention to how social actors understand and respond to educational phenomena, is requisite for the field. Qualitative approaches offer the epistemological advantage of showing how systems, structures, or processes play out "on the ground." Crossley and Vulliamy (1984) criticize CIE scholars for too often presuming that what is known about one context can be assumed to be true in another. This issue, which they refer to as "ecological validity" (198), highlights the importance of examining how cultural, economic, historical, and political forces impinge on schooling within a given context (see also Vulliamy 1990). As Broadfoot (1999) argues, "[E]ducation can only be fully understood in terms of the context in which it is taking place.... The unique contribution of comparative studies is that of providing for a more systematic and theorized understanding of the relationship *between* context and process, structure and action" (225–226).

The absence of contextual knowledge often leads to unsubstantiated and misguided policy prescriptions (Vulliamy 2004). The lack of ecological validity leads to presumptions such as the belief that intercultural education promotes the social inclusion of minority students while advancing economic and social mobility and national unity (see chapters 7, 8, and 9); that standardized tests or supplemental educational services will bridge the achievement gap (see chapters 1 and 6); that teachers enact policies and programs as written (see chapters 3, 8, and 12); that political participation leads to enhanced social capital and educational opportunities in marginalized communities (see chapters 2, 4, and 5); or that educational programs by their very nature reduce conflict (see chapters 10, 11, and 12). Ethnographically oriented case studies have demonstrated that such educational outcomes depend largely on how such policies and programs are perceived and received, as well as on political, social, cultural, and economic constraints. However, extending the work of previous researchers, we argue that contemporary qualitative comparisons require scholars to work *vertically* and *horizontally*.

Comparing Vertically

Vertical comparisons across levels, such as the local, the national, and the international, are essential, we argue, in contemporary qualitative research in the field of CIE. As Marcus (1998) complains, too often qualitative work reifies social, political, and economic processes as "forces" or "systems" with explanatory power. When applied to critical ethnographic studies in CIE, there has been a tendency to take the macro for granted and focus exclusively on a single-site locality rather than carefully exploring how changes in national and international institutions, discourses, and policies are influencing social practice at the school level. For example, the richness of Willis' local-level ethnography was not matched by fieldwork at the British Ministry of Labor or the Organization for Economic Co-operation and Development, a move that might have shown how the resistance displayed by the youth in his study was a reaction to a specific set of conditions of British capitalism during 1970s rather than a more generic outcome of class struggle in any capitalist order. This weakness was, in turn, adopted by many of those doing qualitative case studies in CIE, whereby attention to the local was not matched by ethnographic exploration of the national or international levels.

In contrast, we aver that attention to the ways global processes are shaped by and in turn influence social action in various locales is not optional but obligatory to generate rich comparative knowledge. "The local" cannot be divorced from national and transnational forces but neither can it be conceptualized as determined by these forces (Piot 1999). Thus, we contend that contemporary comparative education research requires carefully tracing vertical relationships across local, national, and international levels (Marcus 1998).

Within the field of CIE, the multilevel analysis put forth by Bray and Thomas (1995) provides an important antecedent to this work. They contend that all comparative education research occurs along three dimensions: the geographic/locational, tied to levels (world region, country, state/province, district, school, classroom, and individual); the demographic not tied to a place (e.g., transnational ethnic or religious groups); and the societal, by which schooling is linked to broader political or economic structures and forces. Instead of comparison across nation-states, Bray and Thomas exhort scholars to compare different dimensions and, specifically, to compare geographic/locational levels to avoid "incomplete and unbalanced perspectives on educational studies" (472).

We extend these insights to argue that contemporary qualitative work in CIE must examine how the global and the local mutually shape one another. Such an approach heeds Tsing's conceptualization of an "ethnography of global connections" (2005, 1). As Tsing writes, seemingly universalizing systems such as capitalism and democracy "can only be charged and enacted in the sticky materiality of practical encounters" (3); they must operate materially in the world. So-called global forces are themselves "congeries" of local-global interactions, gaining momentum or suffering inertia when made materially manifest. To illustrate the study of global connections, Tsing introduces the metaphor of friction. She writes,

> A wheel turns because of its encounter with the surface of the road; spinning in the air it goes nowhere. Rubbing two sticks together produces heat and light; one stick alone is just a stick. As a metaphorical image, friction reminds us that heterogeneous and unequal encounters can lead to new arrangements of culture and power. (5)

Friction, produced through continuous social interaction among actors at various levels, is required to "keep global power in motion," though it may just as easily "slow things down" (6). This metaphor encapsulates "the awkward, unequal, unstable, and creative qualities of interconnection across difference" (6). Global encounters, when conceptualized in this way, often result in new and unanticipated cultural and political forms that exclude as well as enable.

Comparatively knowing, then, requires simultaneous attention to multiple levels, including (at least) international, national, and local ones, and careful study of flows of influence, ideas, and actions through these levels. Qualitative research on education must consider the profound changes in the global economy and inter/national politics that make the national and international levels of analysis as important as the local. Transnational studies in education emphasize the ascendance of supranational policies and institutions, such as the General Agreement on Tariffs and Trade, the World Bank, the International Monetary Fund, and multinational corporations. The growing interconnections between national economies and international financial institutions, and between national educational systems and global organizations that fund and evaluate their operations, are some of the most important issues for scholars of education today. Moreover, the proliferation of commodities identified with the West but often produced in the South calls for comparative methods to understand how communities everywhere adapt to the presence of global cultural and material flows (Appadurai 1996). However, such discussions about globalization in the broader social sciences too often fail to attend to the fundamental work done by schooling in these processes: While scholars might mention schools, they "ignore or downplay the role of modern schools in structuring identities and power relations, both locally and globally" (Levinson 1999, 594). Yet the time-space "compression" (Harvey 1989) instigated by changes in the economy, politics, and technology also requires broader conceptualizations of education, necessitating an approach to local-level research whose boundaries encompass fieldwork sites hundreds or thousands of miles away as well as careful consideration of history and political economy.

Further, vertical case studies de-center the nation-state from its privileged position as the fundamental entity in comparative research to one of several important units of analysis. As Marginson and Mollis (2001) write, "Governance remains national in form, and nation-states continue to be central players in a globalizing world, but partly as local agents of global forces, [as] the nation-state now operates within global economic constraints" (601; see also Dale 2000). Thus, the importance of multilevel research that situates the nation-state within a world marked by global agencies and agendas is more apparent today than ever before. Yet the national-global relationship is only one part of a vertical case study because the local-national and the local-global connections are equally significant. The goal of vertical studies is to develop a thorough understanding of the particular at each level and to analyze how these understandings produce similar and different interpretations of the policy, problem, or phenomenon under study.

The authors in this collection have conceptualized their research as a *vertical case study*, a multisited, qualitative case study that traces the linkages among local, national, and international forces and institutions that together shape and are shaped by education in a particular locale. For example, several authors in this volume use the term *inter/national* to draw attention to the difficulty in separating "national" policy and practice in many countries from the "international" institutions that fund or provide other support to federal institutions (see Vavrus 2005). Further, many of the chapters emphasize the importance of situating interactions across such levels in the context of historical struggles and contemporary cultural politics (see Bartlett 2009). In other chapters, the authors draw on Williams's (1983) notion of *keywords* to contrast "ways of seeing culture and society" evident in the ways different groups of policy actors define and enact critical terms in international development and education, such as *participation* (see chapters 4, 5, and 6) and *sustainability* (see chapter 10). Thus, the authors explore vertical relationships in multiple ways. They demonstrate how historical perspectives illuminate contemporary debates over language policy in Lebanon (see chapter 12) and Peru (see chapter 8); how political economic contexts shape the management of cultural and linguistic diversity in Ireland (see chapter 7), Peru (see chapter 8), and the United States (see chapter 9); how international organizations sway the production and consumption of sexual reproductive health materials in Tanzania (see chapter 3), the conduct of educational reforms in Senegal (see chapter 2) and Tanzania (see chapter 4), and the sustainability of educational programs in post-conflict settings such as Angola (see chapter 10) and Liberia (see chapter 11); and how school-based actors interpret and enact city, state, and national policy in the United States (see chapters 1 and 6) and Brazil (see chapter 5).

Comparing Horizontally

The kind of qualitative research we are recommending does not only study across time and across levels. It also endeavors to study *horizontally* across sites through multisited research in places that are "simultaneously and complexly connected" (Tsing 2005, 6). That is, these investigations "study through" (Reinhold 1994, 21); they examine, ethnographically, "interactions (and disjunctions) between different sites or levels" (Shore and Wright 1997, 14). To do so, the authors draw on three key sources: the concept of *bricolage;* the comparative, sociocultural analysis of policy; and actor-network theory.

The concept of *bricolage,* introduced to the social sciences by Levi-Strauss (1966), describes the creative, improvisational work of interpreting social symbols to respond to ongoing social life. In the field of education,

Ball has drawn upon the concept to describe the process of policy formation. He defines it as

> borrowing or copying bits and pieces of ideas from elsewhere... amending locally tried and tested approaches, cannibalizing theories, research, trends, and fashions and not infrequently flailing about for anything at all that looks as though it might work. Most policies are ramshackle, compromise, hit or miss affairs, that are reworked, tinkered with, nuanced and inflected through complex processes of influence, text production, dissemination, and ultimately recreation in contexts of practice. (2006, 75)

Policy formation is rarely a linear, logical process; a host of political and economic factors shape how policies get made, imported, exported, adapted, and indigenized (see, e.g., Steiner-Khamsi 2004). This improvisational process we refer to as *bricolage*, and several of the chapters (see especially chapter 2) consider the complex cultural politics of policy formation.

Further, we draw upon the sociocultural study of policy to expand the definition of policy studies and to highlight the important social work done with and through these authoritative cultural texts. Levinson and Sutton conceptualize policy "as a complex social practice, an ongoing process of normative cultural production constituted by diverse actors across diverse social and institutional contexts" (2001, 1). Scholars who consider "policy as practice" examine the ways that "individuals, and groups, engage in situated behaviors that are both constrained and enabled by existing structures, but which allow the person to exercise agency in the emerging situation" (Levinson and Sutton 2001, 3). The authors in this volume employ a sociocultural lens to consider not only educational policy (as text or practice) but also educational programs and pedagogies more broadly conceived.

Finally, our efforts to "study through" educational policies and programs are influenced by the framework that actor-network theory provides to examine the multilinear and unpredictable (but often simultaneous and contradictory) *appropriation* by various social actors in various locations of educational policy, programs, and pedagogies (see Latour 1995, 2005). Innovating upon the concept of *bricolage*, appropriation describes the process through which different actors creatively improvise when making, applying, resisting, recasting, and living educational policies, programs, and pedagogies. Drawing on Latour, Koyama explains in her chapter how the theory affords the conceptual tools to follow the "continuous connections leading from one local interaction to the other places, times, and agencies" that are brought to bear on the interaction (Latour 2005, 173). Such interactions create a temporary network of actors (including object-actors, such as policy texts) who form associations and, through their actions, enable and constrain, however temporarily, further actions.

Yet the "horizontal" approach we advocate does not only entail studying "through." Recognizing the commitment of time, energy, and resources necessary to conduct a thorough vertical case study, our approach advocates horizontal comparison through the juxtaposition of cases that follow the same logic to address topics of common concern, as in the case of the chapters presented here. The authors' common frame of reference was not imposed after fieldwork for the purposes of this volume; rather, the similar—though not identical—research design and methodological approach was the result of a specific set of circumstances that brought together graduate students and faculty in the Program in Comparative and International Educational Development at Teachers College, Columbia University during a period spanning nearly a decade (2000–2008). From hours of reading, discussing, debating, and designing doctoral and post-doctoral research projects together, a common framework emerged that allowed each of the contributors to explore her particular topic at multiple levels. This approach is similar to the collection of multisited ethnographies found in *Global Ethnographies* (2000), which resulted from a doctoral cohort working with sociologist Michael Burawoy at the University of California, Berkeley. We, too, became intimately familiar with each of our students' projects because we collaboratively advised them as they were designing, carrying out, and writing up their dissertations. *Critical Approaches*, therefore, contributes to the field of education by providing rich vertical cases and by edifying horizontal comparisons of educational policies, programs, and practices across the countries represented in the collection.

Comparatively Knowing: Strategies for a Vertical Case Study

The studies in this volume are the result of researchers having made two critical political-epistemological decisions: first, deciding that their knowledge of a region was sufficient for the task at hand; and second, determining that the research parameters for the project would permit the development of sufficient understandings of the particular in relation to the national and the global. We believe that scholars using qualitative case study approaches should prioritize the pursuit of historical, cultural, linguistic, political, and economic knowledge that comes with in-depth area studies of a region. Short-term studies by development workers without sufficient knowledge in these areas tend to lack both theory and substance because the conditions of the assignment do not allow such depth of engagement at the local, national, and international levels (Parpart 1995). As Samoff (1999) warns, research commissioned by international develop- ment agencies or other interested parties often promotes an understanding of education and development that comes to be viewed as a global con- sensus. Vertical case study research, we contend, is a particularly effective way to promote engagement with the knowledge borne by various groups of stakeholders in a policy reform initiative, a teacher education program, or a research project (see, e.g., Dyer et al. 2002). Yet, as noted throughout this introduction, the study cannot restrict itself to the local or even the regional; instead, we suggest that qualitative researchers in CIE must be prepared to examine phenomena across multiple levels and numerous sites, even as they also endeavor to situate case studies historically.

The vertical case study approach we are recommending draws upon a portfolio of methods to meet these goals. Here we mention chapters that exemplify the use of specific techniques, though it is worth noting that many of the chapters draw upon several methods simultaneously. The studies might examine archival materials to trace the history of a specific intervention (see chapters 2 and 8); they could consider how geopoliti- cal events shape local conditions (see chapters 7 and 9); and they might employ survey methods (see chapters 11 and 12) or discourse analysis of policy documents (see chapter 7). Many of the contributors to this volume rely upon multisited ethnographic techniques, including interviews and observations (see chapters 1, 5, and 6). Indeed, several entail what might be considered "nested" case studies, in which the scholars research how pol- icies or programs flow between inter/national NGOs and ministries (see chapter 10), local schools (see chapter 3), or communities (see chapter 4) and are, in the process, remade. Obviously, not every study can or should incorporate all of these techniques, but we see them as powerful tools for comparative and international education research.

The chapters in this volume are arranged into four sections that feature key issues in the field. The first section, *Appropriating Educational Policies and Programs*, examines the creative improvisation in which various actors engage as they make, apply, resist, and recast educational policies, pro- grams, and pedagogies. Koyama's chapter employs actor-network theory to consider how various groups appropriate the little-examined Supplemental Educational Services provision of No Child Left Behind in the United States. Max uses the concepts of *décalage* and *bricolage* to consider how a higher education reform financed by the World Bank, backed by the Senegalese government and grounded in participatory consultations, failed to achieve its stated goals. Muro Phillips investigates how inter/national health education materials are appropriated by local teachers in Tanzania whose views on sexuality and sex education for Tanzanian youth differ significantly from those of the staff who develop health education materials. Each of these chapters traces the critical recasting of educational policies, programs, and pedagogies, including a consideration of how actors from international development organizations and inter/national organizations participate in this process.

The second section, *Exploring Participation in Inter/national Development Discourse*, investigates the paradox of the *participation* keyword in development discourse. Although contemporary development projects mandate deliberation by previously excluded social sectors, the first two chapters in this section question the extent to which participation significantly changes business as usual. Taylor shows how the involvement of ordinary citizens in Tanzania's primary education reform known as PEDP was delimited in ways that did not permit participants' priorities to influence policy decisions. In the subsequent chapter, Wilkinson describes how the school councils that originated as a participatory policy experiment in Porto Alegre, Brazil reproduced the historical authoritarian and clientelist relationships they were meant to disrupt. In contrast to these two cases, Hantzopoulos shows how a group of teachers in New York formed teacher-centered, participatory networks to resist the hierarchical, centralized educational reforms that threatened key elements of their pedagogical and curricular models. As a group, then, the chapters in this section consider the limitations of policy and program appropriation, that is, the conditions under which formalized avenues of participation allow for fundamental reshaping of policies and programs.

The third section, *Examining the Political Economy of Diversity*, considers how global political economic forces influence states' efforts to manage diversity. Building on the previous section, it interrogates *interculturalism* or *multiculturalism* as they are currently used in educational development discourse. Working in Ireland, Bryan demonstrates how the political economic conditions responsible for increased immigration, shifting demographics, increasingly unequal distribution of wealth, and intensified racism are precisely those factors ignored in intercultural policy documents and guidelines that purport to alleviate racism in Irish society and schools. Valdiviezo's research in Peru offers interesting parallels. She considers the contradictions of a bilingual, intercultural education (BIE) policy that aimed to combat the marginalization of indigenous language, culture, and ethnicity by developing an ill-informed linguistic and cultural educational policy directed solely at indigenous populations. Nevertheless, Valdiviezo emphasizes how teachers as policy actors have appropriated BIE policy, in ways simultaneously creative and circumscribed, to address the challenges of diversity. Finally, Ghaffar-Kucher's chapter offers an instructive example of how the global political economy, of which the "War on Terror" is a part, contributes to the ways in which Pakistani-American youth develop their own notions of citizenship and national belonging within a school in New York. She argues that the "religification" of urban, working-class Pakistani-American youth, that is, the ascription and co-option of a religious identity, trumps other forms of categorization, such as race and ethnicity, in ways that have significant consequences for students' academic engagement.

The fourth section, *Managing Conflict through Inter/national Development Education*, asks how educational policy and programming both interrupt and in some ways perpetuate conflict in so-called post-conflict situations. Utilizing fieldwork ranging from international organizations to households, these chapters highlight the multiple, interconnected factors shaping decisions about education and schooling. Taking on another important development keyword—*sustainability*—Mendenhall examines the challenges nongovernmental organizations face in sustaining educational support to the fragile post-conflict state of Angola. Drawing on the metaphors of friction and *bricolage*, she shows how the interactions among key policy actors changed during the relief-to-development transition period and how the resulting post-war educational program under study was an improvisation rather than a wholesale transfer of an existing program. Building upon these metaphors and themes, Shriberg considers how international organizations have conditioned key issues of educational policy in post-conflict Liberia, specifically teacher compensation policies, in ways that deeply affect teachers' well-being. The resulting friction produced by "uneven encounters" among international, national, and local policy actors creates new forms of suffering for Liberian teachers while it also perpetuates historical ethnopolitical cleavages in the country (Tsing 2005, 1). The final chapter by Zakharia takes up this concern with how education mediates conflict as it simultaneously returns to questions about the political economy of diversity. Working in Lebanon, Zakharia considers how the Arabic language, made central to post-civil war national unity, is both promoted and undermined by school, state, regional, and global actors during periods of regional and national instability and violence. This chapter vividly illustrates the power of an "ethnography of global connections" to, first, depict the significant role that domestic and regional conflict play in educational policy formation and identity assertion, and, second, to chronicle the new cultural and political forms that arise from encounters among the global, national, and local.

Together, the 12 chapters that comprise *Critical Approaches* demonstrate an approach to CIE ideally suited to the study of the interconnections among actors and object-actors across time and space. Each of the four sections shows how this can be done by exploring a common issue in multiple settings, resulting in rich vertical and horizontal comparisons. It is our intention in bringing together these qualitative case studies that they will expand the boundaries of the field and its customary ways of knowing.

Notes

1. For a similar critique in the British context, see Broadfoot 1999 and Vulliamy 2004.
2. See, e.g., Clifford and Marcus 1986; Marcus and Fisher 1986; St. Pierre and Pillow 2000.
3. See, e.g., Ninnes and Mehta 2004; Ninnes and Burnett 2003; Dale 2003.

Part 1

Appropriating Educational Policies and Programs

Chapter 1

Localizing No Child Left Behind
Supplemental Educational Services (SES) in New York City

Jill P. Koyama

Educational policy can best be conceptualized as a productive social practice. Levinson and Sutton (2001) suggest that policy is "an ongoing process of normative cultural production constituted by diverse actors across diverse social and institutional contexts" (1). As it becomes articulated, policy stimulates and channels actions through levels of governmental organizations, educational agencies, and emerging social structures. To study educational policy vertically is to capture the ways in which policy both constrains and enables localized activities and actions; adapting Tsing (2005), one could argue that policies articulate "heterogeneous and unequal encounters [that] can lead to new arrangements of culture and power" (5). In this volume, inter/national educational policies are shown to empower and disempower local actors (see chapters 5, 6, 9, and 10) while simultaneously providing legitimacy to national governments and inter/national organizations (see chapters 4, 7, and 11). Studying policy vertically diminishes the gap (or what Max—see chapter 2—calls the *décalage*) between official and everyday actions and knowledges to reveal policy actors struggling, negotiating, and acting in ways that constrain or disable policy (see also chapters 3, 8, and 12).[1] Likewise, this chapter traces the diffused and often conflicting actions prompted by No Child Left Behind (NCLB), the federal educational policy of the United States, in relation to its Supplemental Educational Services (SES) provisions.

In American education, federal policy has long served to politically regulate the efficiency of a schooling system that operates on multiple legislative and organizational levels and to set normative rationales for the actions of the state. Schools and other institutions are obliged to follow federally generated policy. However, such policy is not an "intrinsically technical, rational, action-oriented instrument that decision makers use to solve problems and affect change" (Shore and Wright 1997, 5). Rather, as demonstrated in this chapter, educational policy is a process of *bricolage*: Policy is made, negotiated, resisted, and remade by the actions of multiple agents across multiple local and delocalized situations (Ball 2006).

Federally mandated, state-regulated, district-administered, and school-applied, NCLB touches nearly every facet of American schooling. At schools, it changes the daily routines of millions of students, parents, teachers, and school administrators. More broadly, it connects the actions of numerous agents in multiple institutions, that is, assessment development companies, colleges of teacher education, and tutoring companies, which are associated with schools but have rarely been recognized for their integral roles in public education. Vertically studying the appropriation of NCLB—the process by which different actors creatively apply particular elements of the policy to their situations—brings the associations between public and private entities to the fore, placing the organizational and everyday worlds on equal footing by blurring the boundaries between government, schooling, and commerce.

In this chapter, I provide in-depth views into the previously understudied relationships between schools and for-profit educational services. I follow the appropriation of NCLB—particularly its supplemental educational services provisions—by tracing the dynamic linkages between the New York City school district, public schools, city government, and United Education, a for-profit SES provider. This study demands a shift in the analytical focus from students or teachers (as necessitated, e.g., in chapters 3, 6, and 9) to situations that occur outside of classroom instruction and beyond official school hours, including public settings such as SES provider fairs, publicized Department of Education (DOE) events, school district meetings, parent teacher association meetings, and after-school tutoring.[2]

My research uses a multisited ethnographic approach to study through, vertically and horizontally, the situations afforded by the SES educational policy. I explore the ways in which many actors, from diverse institutions and organizations, incorporate elements of NCLB into their particular situations and examine how their actions hold them temporarily, and sometimes tenuously, together. Utilizing actor-network theory (defined in the following text), I trace the string of actions through which adults, in and out of schools, attempt to remedy failure, the goal of NCLB. Their actions represent "the agency of local actors in interpreting and adapting such policy [NCLB] to the situated logic in their contexts of everyday practice" (Levinson and Sutton 2001, 17). Although NCLB presupposes that actions directed at school failure, including SES, can be comprehensively planned and choreographed, the actors in this study demonstrate that attending to the "problem" does not necessarily require consensus in motivations, objectives, or practices. Following them demonstrates how their activities help to further construct the problem of school failure by leaving behind some of the very students at which the policy is aimed.

Federal Educational Policy: No Child Left Behind

> No Child Left Behind is a big government, big political solution to all the things, like students who can't read and teachers who can't teach, that I deal with everyday here.... On paper, No Child [Left Behind] looks like some big great deal. At our level here, what I call "the real and regular," No Child doesn't solve as many problems as it makes. You'll see. It won't solve low achievement either. (Interview with Queens middle school principal, November 7, 2006)

In the United States, the federal NCLB policy extends the historic promise made in the 1965 Elementary and Secondary Education Act (ESEA), which stipulated that all children, regardless of race, socioeconomic status, gender, creed, color, or disability, will have equal access to an education that allows them to enjoy the freedoms and exercise the responsibilities of citizenship in our democracy (Wood 2004). From a historical perspective, NCLB is not an abrupt shift in U.S. educational policy. It is a policy made possible, if not probable, by nearly two centuries of schooling in America, in which individual meritocracy, accountability, and standards—cornerstones of NCLB—have been made increasingly prominent through policy and practice.

However, some elements of the policy are unparalleled. Since its passage in 2001, NCLB has become central to the organization and governance of America's educational system. The legislation's 588 regulations focus on improving student achievement and helping close achievement gaps in the nation's nearly 99,000 public schools by radically centralizing the standards of learning and assessment. Pushing substantial organizing and regulatory power toward the federal government through unprecedented assessment requirements and student performance goals, NCLB changes the meaning of the 1965 Act and alters the actions required by those in local administrations, schools, and educational support companies who find their daily lives impacted by it.

Under NCLB, schools must improve the academic achievement of students. Schools that repeatedly report insufficient progress are designated as "schools in need of improvement" (SINI), NCLB's nomenclature for "failing," and they are deemed incapable of improving through their own efforts. Needing improvement for three continuous years requires schools to use federal money to contract companies to provide SES to students who are receiving free lunch. Eligibility hinges upon the child's free-lunch status, rather than his/her measured academic achievement. Ironically, a child who has consistently received low marks, but who qualifies for reduced, not free, lunch cannot receive SES; conversely, a child who consistently receives high marks and receives free lunch can receive SES.

The SES provisions legitimize the need for resources beyond those regularly available through the school system by federally authorizing companies outside of public education to teach children directly through curricula and programs developed independently but paid for by Title I funds—funds that the federal government has allocated to the country's highest-poverty schools to improve the academic achievement of the lowest-performing students since the initial passage of the ESEA in 1965. The funds used for SES are reallocated from the Title I funds, which were already granted to schools but were slightly increased under NCLB; they do not represent separate additional funds for school districts or individual schools (Sunderman 2007). In fact, administration costs, such as parent outreach, cannot be paid from the Title I set asides, and it is estimated that NCLB costs states nearly 10 times as much as they receive from the federal government.[3] According to NCLB, school districts must set aside 20 percent of their Title I funds to cover the cost of providing SES and the expense of transferring students out of "failing" schools. In New York City, transferring to a more "successful" school, an option known as "public school choice," is not encouraged by the practices of the DOE; instead, the majority of funds are used to pay SES providers approximately 2,000 dollars per child for 100 percent attendance in a tutoring program.

Paradoxically, the partnerships between schools and SES providers exist within an increasingly centralized governance of public schools and a competitive free market. The SES provisions of NCLB explicitly expand the role of the private sector in public education begun 25 years ago by "A Nation at Risk" (National Commission on Excellence in Education 1983), a report that demanded school reform based on the notion that failing schools could be improved by corporate expertise. Thus, SES is not necessarily a "legitimate representation of 'public' needs and interests" (Levinson and Sutton 2001, 2), but instead a sociocultural process that mandates schools to enlist services and products provided from organizations and companies outside of schools in their efforts to eliminate school failure.

Theoretical Perspective: Actor-Network Theory

I draw upon actor-network theory (Latour 1995, 2005) to examine the dynamic work and the relational links set in motion by the appropriation—commonly referred to as the formation and implementation—of NCLB and "to grasp the interactions (and disjunctions) between different sites or levels in policy processes" (Shore and Wright 1997, 14). The theory provides a way in which to follow the "continuous connections leading from one local interaction to the other places, times, and agencies" (Latour 2005, 173). From the federal government, through state and district educational agencies, to individual schools, the SES provisions of NCLB are implemented through the actions of many. Policy is not imposed in a linear fashion by the federal government onto schools. Rather, the appropriation of provisions such as SES progresses though multiple vertical and horizontal interactions between private, school, state, district, and federal entities. To understand educational policy, we must examine these social interactions.

The interactions create linkages that develop into a network. The construction of the network begins with a situation that is deemed in need of solving. In this chapter, the situation, as posed by the federal government, is that the SES provisions of NCLB will help eliminate school failure for children. To legitimize SES's ability to increase the academic achievement of students (as measured by increased scores on standardized tests), the federal government must enroll other participants. During a period that Latour (2005) refers to as "a period of problematization," the primary actor (NCLB) finds relevant actors (SES providers, failing schools) and delegates representatives from groups of actors (principals and tutoring managers) into roles. The federal DOE uses multiple strategies to get participation from the actors; the most basic strategy in this case relies on mandating state and district participants that "no child should be left behind" by 2014. Another strategy, selectively presented to principals, encourages the schools to participate robustly in SES so that they can get off the failing schools list, be rid of the outside providers, and thus begin to implement their own programs. Actors, such as principals and SES providers, variably invest in their roles. Finally, a differentiated aligning of roles begins to emerge; actors mobilize, form associations, and construct their environments. They do what they can or must do to implement SES; in the process, other situations arise and the process begins again.

The cycle described earlier is referred to in actor-network theory as "translation," or the interpretation given by participants or "fact builders" of their interests and that of the people they enroll (Latour 1987).[4] Through translation, actors associate with other actors to form joint vectors of agency. Translation is driven neither solely by the agency of an actor nor by "larger" forces. In this study, administrators of failing schools and the managers of for-profit tutoring companies—who may not share aims, explicit interests, intentions, or regulations—form relationships that depend on multiple and repeated translations.

Further, the human actors do not act alone. Objects with subjective investments also become actors. The nonhuman actors in this study emerge when experiences are transcribed into artifacts bearing the imprint of their creators and then into fact (a discovery that comes to be accepted by the collective as established and often no longer controversial) (Taylor and Van Every 2000). Latour (2005) focuses on the transcription of findings, or production and validation of cultural texts like policies, in the context of social collectives. Policies, as object-actors, help human actors make some sense of how the federal mandates apply to them. By requiring specific kinds of actions, the trajectory of objects is traceable. Notably, in an emerging network, "non-local" actors (such as the federal government) become localized through object-actors (such as NCLB policy, district reports, or standardized test scores) that circulate, assisting actors in the construction of competence to act. As actors and materials are sent from one local place to some other place, that which was global becomes part of the network and thus becomes vertically localized. Actor-objects play an important role in producing the "friction" of global encounters (Tsing 2005).

The Study: Methodology and Entities

The findings I present here on SES emerged from a larger multisited ethnographic study in which I focused on the actions between NCLB, New York City's governing educational bodies, and 42 schools that entered into partnerships with for-profit tutoring companies for 3 years, from June 2005 to June 2008. The company discussed here, dubbed United Education, is a composite company that I assembled from the data of five similarly organized for-profit national SES providers that offered services in New York City; the composite is necessary to protect the privacy of participants in the study. The companies share comparable bureaucratic organization, as well as analogous employee titles and positions. Their services and structure are so similar that in several government reports and in media releases the practices of the individual companies were lumped together, under the rubric of "well-known" or "well-established" SES providers.

Data for the study were compiled from many sources: interviews with principals, assistant principals, parent coordinators, and other school staff; interviews with members of the DOE's regional superintendent offices, the Office of School Support Services, the Office of Strategic Partnerships, the Mayor's Office, the Middle School Task Force, the Education Committee of the City Council, the city's Youth Services, and a variety of other boards and panels associated with NCLB and Children First reforms in New York City; interviews with United Education lawyers, managers, directors, product developers, curriculum writers, teachers, and sales people; observation and participation in afterschool SES programs, governmental meetings, DOE meetings and seminars, school meetings, teacher-training sessions, community assemblies, policy forums, and United meetings; and collections of NCLB regulations, SES materials, school documents, and pieces of diverse public media.

This chapter focuses precisely on the continually fluctuating actions and associations between multiply-situated actors. The actions served as ideal "tracers" toward an array of connections within an emerging NCLB-SES actor network. I followed these conduits—people, materials, and documents—as they circulated NCLB-directed action from one place to another, secured temporary connections, and over time constructed a traceable actor-network that extended the field of study beyond schools. I took the "field" to be more than a set of physical locations; it represented metaphorical spaces, transactional places constructed by social interaction, and literal spaces and local places where actors acted.

In a day of fieldwork, my inquiry led me to participate at a morning SES provider fair, a parent association luncheon, an actual SES program within a school, and finally to an evening town hall meeting for parents disgruntled with the city's educational reforms. Other days were spent pouring over NCLB policy and SES provider regulations, interviewing principals and regional superintendents, and attending public forums sponsored by the DOE. Much time was spent physically following actions across the linkages between levels and organization of two multiply-situated entities—New York City public schools and United Education.

In 2003, informed by NCLB directives, Mayor Bloomberg and Schools Chancellor Klein launched the initial phase of the Children First Reforms across New York City's public schools.[5] The first phase of the reforms, referred to by Bloomberg as the "unifying and stabilizing period," centralized authority by instilling a standard curriculum and reorganizing the city's 32 districts into 10 bureaucratic regions. All actions, according to the reforms, were to center on three core principles—leadership, empowerment, and accountability. In 2007, the second phase of the reforms—essentially a decentralization of authority and power—was implemented. The ten regions were abolished, principals were given greater autonomy in staffing and budgets, and schools were offered three "support organization models" designed to help them reach their accountability targets, provide professional development, ensure high-quality teaching, and design programs to increase students' test scores.[6] Both phases of Children First Reforms were appropriated simultaneously with the NCLB mandates, which, among other directives, required failing schools to seek services from SES providers.

SES providers across the United States are quite diverse. In 2004, 69 percent of the providers were private, 25 percent were school districts, 2 percent represented college or university programs, and 4 percent had unknown affiliation (Sunderman 2007). Some represent large test preparation corporations, such as United Education, that are recognizable by name and that boast multiple decades of practice providing academic assistance to students, while others are small local organizations. By 2005, across the country, there were 1,800 registered SES providers. The same year, the SES providers were poised to earn nearly US$200 million, with the large for-profit national companies, such as Kaplan K12 Learning, The Princeton Review, and Sylvan, securing one-third of the profits.[7] In New York City, the number of approved SES providers increased from 47 in 2002–2003 to 174 by 2006–2007 (Sunderman 2007).

In my work, I discuss a composite company, United Education. The actions, practices, and products I attribute to United Education are drawn from five actual companies, and the data I present are drawn from situations and activities I observed. Much of the information pertaining to United was obtained from public documents and widely circulated and published communications. Most of the SES data were available through electronic links from the Web sites of federal, state, and district educational agencies. Material that was garnered through interviews, personal correspondence, staff meetings, or internal documents has been paraphrased or summarized to eliminate identifying information and to protect the confidentiality of informants.

Findings: Nonlinear and Conflicted Actions

Within the emerging SES actor-network, the DOE, the schools (namely, administrators and parent coordinators in this chapter), and the SES providers worked side-by-side, and sometimes together, to adapt NCLB;

however, each group's aims and circumstances resulted in variable and often disparate implementation of the SES mandates. The DOE and the "failing schools" within the district were mandated by federal law to employ SES programs. Noncompliance would result in the loss of federal educational funds, and the inability to reach the NCLB goals would result, at least for schools, in sanctions. In contrast, the SES providers chose to assume their roles and stood to gain substantial amounts of federal money regardless of whether schools met their NCLB goals. As shown throughout this section, the three actors—the DOE, schools, and SES providers—were often at odds with each other. They played their network roles according to the aims and requirements of their situations, not necessarily to NCLB's goal of attending to all children. This was evident in an interview with a Brooklyn elementary school principal, who described the relationship between schools like his and SES providers:

> Basically, the feds say that we can't teach kids as well as companies can. NCLB tells us that we must teach ready-packaged curriculum all day and then get some company like Princeton Review or Sylvan that uses their own curriculum to run *our* afterschool programs like we don't know how to do it...And the feds don't think that's going to cause problems. Ha. (Interview, February 6, 2006)

As relative unregulated interlopers to the school system, SES providers were often viewed by other actors as suspect. Although the regulations—and sanctions—of NCLB focus on accountability for children and schools, the policy provides no equivalent requirements for these providers. The SES providers are exempt from meeting the highly qualified instructor requirement of NCLB, are exempt from offering services for students designated as "disabled" and "limited" in their English proficiency, and are exempt from adopting a standard curriculum. There is little oversight over the curriculum, lessons, and evaluative measures used by SES. Each program evaluates and reports the progress of its own students according to its own assessments. In the words of an assistant principal of an elementary school in Manhattan, "They [the SES providers] don't have to do anything by the book because they don't have a book!" Indeed, while the procedural requirements, such as submitting attendance and enrolling students, are micro-managed by the DOE, the actual content of SES programs is largely unexamined.

According to NCLB (U.S. Department of Education 2001), the NYC Department of Education, as the city's local educational agency, must ensure that each SES provider has a "demonstrated record of effectiveness in increasing student academic achievement" [Section 1116(*e*)(12)(B)(*i*)] and "uses strategies that are high quality, based upon research, and designed to increase student academic achievement" [Section 1116(e)(12)(C)]. Some, including Patty Sullivan, the director of the Center on Education Policy, a Washington-based research group, argues that SES is not well or consistently regulated by federal, state, or local authorities. Sunderman (2007) concurs, noting that "for providers, the basic requirements are minimal" (4). SES providers' accountability, albeit minimal, is to educational agencies and not to the individual schools or students.

School administrators across the city argued that they should not be required to rely on SES providers to help increase students' achievement. Most administrators in this study took greatest issue with the gap in accountability between themselves and their selected SES providers. A Bronx Middle School assistant principal expressed her doubt and distrust:

> Why should we pay for it [SES] when it isn't even proven? Basically, our school and our teachers are being told we don't know how to turn our kids around. We should just turn it over to SES providers when we have no idea if they are capable or not. Doesn't that sound a bit crazy to you? And then, we will just wait, holding our breaths to see if it works or not. If it does, then great. If it doesn't, we lose our jobs.... Who would do such a thing? We have to because the law [NCLB] tells us to. Hire multi-million dollar companies to bring in all their books and staff and just have our teachers stand by and wish for the best. Crazy.... And of course, we do it because we have to. (Interview, September 28, 2006)

The legitimacy of the principal's doubts about SES effectiveness was supported by Sunderman (2007), who found in her metaanalysis of SES studies that there was no research documenting how effective SES might be for improving student achievement. Aside from a few small and not particularly rigorous studies, conducted primarily at the school and district levels, there exists scant research on the effectiveness of SES.

Many school administrators denounced not only the gap in accountability between SES providers and schools, but also the ways in which the money paid to SES providers drained their school's financial resources. A Brooklyn principal expressed the concerns:

> I could use this money to buy more students a couple of reading coaches or math experts...Face it, schools don't want to give this money to multi-million dollar companies to do some elevated homework help. We know where we need the money and we aren't too happy that the feds are telling us we need to give it to SES providers...So, I'm basically stuck. I've got to provide SES and I have to use some of my money. That means we won't have some programs we need in the day. (Interview, November 13, 2007)

Other principals similarly objected and petitioned the DOE for additional separate funding to replace the funds spent on SES.

Principals often resisted their roles as SES partners by limiting their spending on SES programs. Several administrators who chose United minimized the amount of Title I funds they spent on SES by artificially creating a condition in which the demand for service, as measured by numbers of students enrolled, was reduced. According to NCLB, school districts must spend at least 5 percent of the Title I funds set aside for SES unless demand for services is low. One principal of a Bronx middle school demanded that United Education offer SES only on Saturday despite the company's data showing dramatically lowered participation rates for Saturday programs. He told the United field manager:

> If I have to have you guys here, I'll have you here on the weekend. And you'll have to pay for the permit and security. You know I don't like you, don't like the whole SES thing. You cost us money and I don't think you're worth it...I don't want to see you here during the week. I don't want you marketing during the week and I don't expect you to have a large program. (Fieldnotes, October 22, 2008)

To ensure he would spend minimal amounts of his funds on SES, this principal also limited enrollment by distributing enrollment packets only to students with the lowest test scores. Only 38 students enrolled in the SES program, which experienced a 32 percent attendance rate. He then used the money he saved to pay for a Saturday Academy, which was attended weekly by more than 150 students.

By spring 2007, principals were given a DOE policy-based incentive to further circumvent the SES mandates by directing more resources toward improving school environments and targeting support to students who were on the edge of measured English Language Arts (ELA) and math proficiency. The New York City DOE began issuing each public school an A–F scale progress grade, based on its score in three categories: school environment (15 percent), student performance (30 percent), and student progress (55 percent). While the school performance category represented students' annual standardized test scores, the category of student progress measured schools' year-to-year gains in moving students to proficient levels in ELA and math. School environment ratings were based on the results of surveys taken by parents, students, and teachers, as well as student attendance rates.

Coinciding with the progress reports, principals began simultaneously discouraging enrollment in SES and encouraging students' participation in enrichment programs. A United manager suspected that the principals were attempting to increase the "school environment" scores on the school's annual report cards. Her suspicion was confirmed by a Queens elementary teacher, who explained,

> Well, it's like this. It'll be really hard for us to make too much progress in student performance. We already are doing really well this year in ELA and math. Our kids even scored well, so we need to get points for school environment and what parents and teachers said was that there was too much test prep and not enough enrichment. So, now we have enrichment. What can I say? It's all about the scores which are tied to accountability which are tied to our budget. (Interview, March 1, 2007)

A Bronx principal confirmed the tactic, noting that he was also trying to raise his score by offering more diverse afterschool options. He admitted that he would "reroute" some of the federal Title I funds he was allotted for SES to administer his new programs. According to the Education Industry Association, a lobbying group for 800 corporate and individual SES providers, the principals in this study were not alone. The association accused schools and districts across the country of trying to "dissuade parents from accepting tutoring on grounds that it would eat up federal aid that schools need for other reasons" ("Parents" 2006).

Schools were not the only actors taking actions to limit SES; the DOE, the governing body of America's largest school district, seemingly reduced initial access to SES. While the number of students eligible for SES services in the City never fell below 200,000, the number of them who enrolled in SES programs has decreased every year since 2003–2004.[8] The greatest decrease of 14,000 students was experienced in 2006–2007, prompting United's managers to accuse the DOE of providing an insufficient number of forms and not regulating how they were distributed. During a staff meeting, he lamented,

> So, they [DOE] say they want us to enroll more students, but they aren't giving schools forms. No forms, no students. Low overall enrollment numbers this year. It's an easy equation, in that sense. Schools have asked, right? And they haven't gotten any more [forms], so that's just where we are for now. They don't do their jobs and we lose revenue. (Fieldnotes, March 14, 2007)

Other managers expressed similar frustration during meetings with principals and parent coordinators, but they emphasized that the kids, not United Education's profits, were suffering. Principals and parent coordinators agreed that the DOE's unwillingness to provide and distribute forms was having a negative effect on SES enrollment.

Enrollment forms were given to families at the beginning of each school year. First, DOE officials determined how many SES packets and in what languages each school should receive; second, school administrators determined how to get the packets to the eligible students. Since only students who received free lunch were eligible for services, the school calculated who was receiving free lunch.[9] There were often changes from one year to the next, and the final free-lunch rosters were not compiled until mid to late October, the same time SES programs were scheduled to begin. Thus, every autumn estimates were made when determining how many forms each school would need and sometimes enrollment packages were given by the parent coordinators to families that did not qualify.

In fall 2006, schools began reporting a shortage of enrollment packets. A few schools reported receiving no SES packets, while others received materials in languages incongruent with their student population. Of the thirty-six schools with which United had partnered that year, nearly two-thirds had not received enough forms and three had received forms only in Spanish. Parent coordinators distributed enrollment packets based on various criteria, including grade level, test scores, and class number. Such dispersion systems led to parent outcry. Why had a sister's child received a form instead of her child? When would her form be coming? Why was his child not receiving free tutoring? Had her child done something wrong in school to deserve this omission?

Officials cited several reasons for their refusal to supply additional forms, including a previous waste of unlimited forms and their improper distribution by parent coordinators. According to a regional superintendent in Brooklyn, they were only reigning in an unwieldy system that had previously allocated forms to students who then received free tutoring for which they were not eligible. She explained,

> We need to get them [parent coordinators] to be mindful of who gets forms and who doesn't. Do you know how many calls I get every day from parents who signed their kids up for SES only to be rejected by us because they don't qualify? If they just gave forms only to free-lunch families, we wouldn't have this problem. (Interview, December 7, 2006)

Some parent coordinators did give forms only to those families who qualified for free lunch. However, because the free-lunch determinations were not made until after SES enrollment began, many students who were found to qualify in October and November did not receive enrollment packets by this method. Regional superintendents and DOE officials also accused SES providers of "hijacking forms." One explained,

> Listen. We know there's a problem. I must've gotten ten calls today from PCs (parent coordinators) asking for more forms, but none of them could

explain where missing forms were. I told them if they can give me an explanation for the disappearance, then I'll get them more forms...I can't just be sending more and more forms out there. I don't mean to sound paranoid, but each form represents 2000 dollars for a provider and we have been told that [SES] providers are out there buying enrollment forms. (Interview, December 12, 2006)

When asked for details, she admitted that the claim was unsubstantiated but emphasized that the "higher ups in the Board" [DOE] were taking it seriously.

Despite these claims, neither of the reasons offered by the superintendents for inadequate forms seemed plausible. For a student to be admitted to an SES program, there were a series of steps and checks in place to ensure that a student not only be eligible but also that the student had enrolled in only one program. Each enrollment form required a label, which could only be generated by the DOE database, before being submitted to an SES provider, so providers could not actually "forge" enrollment forms. Because each enrollment form was processed by the DOE, there could not be duplicates. Reporting attendance was also highly centralized to ensure that SES providers billed only for time attended, in quarter-hours.

Although there was parental demand in many schools, the DOE continued to ignore the requests for additional forms. According to a parent coordinator in the Bronx, "I've written. I've called and now the principal has emailed and called—and still no forms. I don't know what it'd take to get some." Her account was echoed by several parent coordinators who claimed that their pleas for more forms had gone unanswered.[10] There was one parent coordinator from a Queens middle school, who, desperate to get forms into the hands of her students, photocopied additional enrollment forms. She had only received enough enrollment packets for two-thirds of the student population and despite numerous calls to the regional office she had not received any additional packets. Yet parents were streaming into her office by the dozens to get enrollment forms after the school's open house.

To solve the problem, the parent coordinator took practical action. She photocopied enrollment forms and distributed the full-colored photocopies that were a good likeness for the originals, although the paper stock was unmatched. Two hundred and twelve students returned the counterfeit forms, which were submitted to United and accepted by the DOE. However, as more parent coordinators photocopied counterfeit versions of enrollment forms, parent coordinators and principals were notified by Kathleen Lawrence, the DOE head of SES, that photocopies would not be accepted. Students who used the counterfeit forms were denied access to the SES programs even when they qualified for the services. Resubmitting a form was often not an option since the schools continued to be without sufficient numbers of DOE-distributed forms. So, even though the actors in this network— the DOE, the parent coordinators as school representatives, and United, an SES provider—were behaving in accordance with their interpretations of SES roles, some children, in fact, were left out of the programs.

Concluding Thoughts

NCLB steers action toward school failure in a generic manner; that is, local educational agencies, schools, and SES providers must do something about it, but what exactly they do is somewhat flexible. Collecting the local agencies, schools, and the tutoring industry together implies a certain regulation of actions. However, as demonstrated in this chapter, it does not prohibit the flow of activity in multiple, and nonlinear, directions. NCLB directs but does not determine action. In fact, many small acts of defiance in everyday actions (de Certeau 1984) were made: Principals reduced the availability of SES, and the DOE implemented a school evaluation system counterproductive to stimulating students' participation in SES. Technically, these acts were a refusal to attend to school failure as prescribed by the SES mandates of NCLB. Still, others, including parent coordinators who created contraband enrollment forms, tried to attend to the goals of NCLB by engaging in activities that placed them in noncompliance with the SES process.

This vertical investigation exposes the ways NCLB "does" things across multiple levels of government, schooling, and commercial entities. NCLB mandates, including SES provisions, are "real things" to which many actors in this study attended, but the policy's agency extends far beyond the actual statements and directives it makes in its 600 pages. Much of what it does is more diffuse—through the way collectives interpret and construct the policy for their own means and by the way agents orient themselves and their actions in relation to the policy, to governing agencies such as the DOE, and to each other. The actors in this chapter were all taking practical actions to make sense of their ceaselessly changing circumstances under NCLB.

Schools have long been recognized as agencies through which educational policies targeted at school failure are articulated; private tutoring and testing companies, such as United Education, have not. NCLB mandates—in particular, those associated with SES—are bringing many more actors attending to failure to the fore. As seen throughout this chapter, SES providers are situated as legitimate educational institutions who attend to failure through everyday actions and activities loosely directed by NCLB. The interactions prescribed by NCLB and examined in this study between a well-established for-profit SES provider and its New York City partner schools reveal that attending to school failure provoked a host of unintended consequences—not the least of which was leaving some children out of the very programs intended for them.

Similar to other studies in this volume, this chapter highlights the participatory nature of educational policy and practices across networks or polities of practice, where actors are both consumers and producers of policy. Studying policy processes vertically highlights the importance of interactional spaces, where many actors—from multiple levels of national governments, inter/national organizations, and local educational institutions—respond to policy, appropriating it to meet their goals. Policy processes and solutions are a *bricolage*, an improvisation from one minute to the next in response to and in concert with the available cultural materials and social relations (see chapter 2). As shown throughout the volume, appropriating policy is political, often deeply embedded in actions of commercialization (see chapter 3), decentralization (see chapter 4), sustainability (see chapter 10), and centralization (this chapter). Examining the complexity of these actions vertically, to simultaneously compare multiple contexts, is essential to the development of comparative education scholarship, which has often too readily accepted contexts as a priori explanations to educational phenomena.

Notes

1. Official policy refers to that generated and circulated in the government and emphasizes current political-operational needs.
2. I use "afterschool" rather than "after-school" in referring to the SES programs. I follow Noam et al. (2002), who state that the term "afterschool" (one word) "conveys the institutional legitimacy of the field rather than a tangential add-on to the institution of school" (18).
3. Overall, federal spending accompanying NCLB represents approximately a 1 percent increase; Mathis (2004) calculated that a 30 percent annual increase in current school spending would be needed to come near to NCLB goals.
4. In Latour's (1987) theory, "translation" hints on its common usage in literacy and also as that of a connection that transports or moves thing from one place to another.
5. For more information on Children First Reforms, see http://schools.nyc.gov/Offices/ChildrenFirst.htm.

6. To learn more about the support organization models, see DOE Press Release ID N-42, 2006–2007.

7. More information on SES in New York City, including a list of providers, can be found at http://schools.nyc.gov/Administration/NCLB/SES/default.htm.

8. In New York City, the number of SES-eligible students fluctuated between 243,249 in 2002–2003 and 208,016 in 2006–2007. The number of enrolled students was nearly 87,000 during the 2003–2004 academic year.

9. Some schools are designated as "universal free lunch sites" for three years at a time. At these schools, all children are eligible to receive SES.

10. In March, after the SES program ended, boxes of enrollment packets arrived at the very schools that had requested them in September.

Chapter 2

The *Décalage* and *Bricolage* of Higher Education Policymaking in an Inter/National System

The Unintended Consequences of Participation in the 1992 Senegalese CNES Reform

Rosemary Max

In the immediate postindependence era, universities in Africa were often described by African scholars as "engines" (Ki-Zerbo 1974) of national development and "the greatest weapon of nation building" (Habte and Wagaw 1993, 680). African governments worked to turn these colonial institutions into national universities that would transform not only the individuals who attended them but also the African nations themselves. The reality today, however, is that many public campuses are chronically underfunded and overcrowded. African faculty members are leaving for jobs abroad with greater remuneration and better working conditions; buildings are in need of renovation; and students' studies are interrupted by strikes as well as prolonged by purposeful decisions to remain on scholarship rather than graduate and face a weak labor market.

Although participants within African universities have discussed these challenges since the 1980s, only recently has the plight of these institutions received international attention.[1] Notably, the World Bank has become the largest single international actor with respect to higher education in Africa.

From 1996 to 2007, it lent an average of US$323 million annually to the higher education sector in developing countries, of which Africa received 13 percent or nearly US$50 million a year.[2] The World Bank cited 55 active higher education projects on its tertiary education web site in 2008, many of which are in Africa. Departing from its approach before 2000, which has been described as "piecemeal" and hampered by a "narrow focus,"[3] today the Bank promotes higher education as a critical link in the development chain fostering the "knowledge economy" (World Bank 2000).

One important precursor to the Bank's sudden attention to higher education was a Bank-funded higher education reform project at the *Université Cheikh Anta Diop* (UCAD) in 1992 in Senegal. The reform, called the *Concertation Nationale sur l'Enseignement Supérieur* (the National Consultation for Higher Education, or CNES for short), stood out in the Bank's portfolio for many reasons. This reform was the first effort of its kind in Africa. The Bank had on rare occasions supported African universities, but nothing compared to the level of support it gave UCAD during the CNES. The reform was also quite comprehensive, and it covered many aspects of higher education, from financing and student services to the purchase of lab equipment and library acquisitions. Moreover, the Bank set aside US$40 million for the project, an amount that dwarfed the UCAD budget at the time. The CNES was also striking because it began at the height of the structural adjustment era. Senegal, like other countries subject to strict lending conditions from the Bank and the International Monetary Fund (IMF) at that time, was directed to shrink the size of its public sector to meet the conditions for obtaining loans. Education spending did not escape this restructuring as the Bank encouraged countries to shift resources from the tertiary to the primary sector. The Bank's reluctance to fund universities was based on rate-of-return analyses by its own economists that found lower returns to investment in higher education when compared to primary and secondary schooling (Psacharopolous 1980). Thus, the Bank concluded that scarce public resources should be redirected to the primary education level, leaving higher education a private, not state, responsibility. However, it was precisely within the context of structural adjustment discussions with the Senegalese government from 1989 to 1992 that the Bank decided to invest heavily in a higher education reform project in Senegal (World Bank 1996).

An examination of this apparent contradiction reveals multiple motivations for the Bank's involvement in higher education in Senegal, ranging from restructuring education spending in line with its neoliberal economic philosophy to quelling university-based political opposition that often took aim at the Bank's involvement in the Senegalese economy. However, the Bank's stated purpose for its involvement in a university reform process was to improve higher education in Senegal through a participatory, consensus-building exercise. Thus the CNES began in 1992 and developed into a 14-month national consultation on higher education. Although there was reluctance on the part of many university faculty and staff members to participate in a government-proposed and Bank-funded process, they eventually came together in a rather extraordinary effort to reform higher education in Senegal. Hundreds of participants representing the university community, government, private business, political parties, and various other sectors of society met in working groups, plenary sessions, and informal discussions under the auspices of the CNES. Although discussions were at times contentious, the CNES eventually produced a 23-point reform agenda that was adopted by the government, ratified by the university council in December 1993, and implemented at the beginning of the new academic year in October 1994.

Despite the length and breadth of the process and the millions of dollars spent, the CNES ended with very few of the reforms actually implemented and with the Bank canceling what it called an "unsatisfactory" project (World Bank 2003, 23). This "failure" raises the compelling question: How did an internationally financed reform backed by the Senegalese government and grounded in participatory consultations with higher education stakeholders from around the country derail? Scholars have long studied external influence on educational systems in African countries and detailed the negative effects of the colonial relationship on education policy (Blakemore 1970; Bouche 1975; Mazrui 1975; Altbach 1978; Brock-Utne 2003). However, scholars have also indicated that there is a need for a deeper understanding of *how* global and local forces continue to intersect in a post-colonial world where international actors still play a role in national policy environments (Samoff 1999; Carnoy and Rhoten 2002; Monkman and Baird 2002; Samoff and Carroll 2004; Vavrus 2005). It is clear that a one-dimensional analysis of international and/or local influence should be discarded in favor of a more nuanced interpretation of the "friction" that emerges when various forces and actors engage with and work for change within a specific social, political, and economic context (Tsing 2005). For this research, a vertical case study methodology provided the space to compare "knowledge claims among different social locations in an attempt to situate local action and interpretation within a broader cultural, historical, and political investigation" (Vavrus and Bartlett 2006, 95). This study explores such knowledge claims by situating higher educational reform historically before comparing claims made by diverse categories of participants in the CNES reform that occurred in the 1990s. Such an approach yielded new insights into why the CNES reform process failed and revealed what I refer to as the *décalage* and *bricolage* of higher education policymaking. *Décalage*, which means, quite literally, a gap, refers in this case study to the difference between stated educational policy and actual practice. Further, I use the concept of *bricolage* to describe the improvisational rigging together of pieces and parts of policies and programs, a process of tinkering toward solutions (see also Ball 1995, 2006; chapter 1, this volume). These concepts—*décalage* and *bricolage*—aid in understanding how higher education policy is made in an inter/national system in which post-colonial governments remain severely constrained despite (and perhaps because of) the participation of formidable local constituencies.

1918 to Independence: *Décalage, Bricolage* and the "Embryo" of Higher Education in West Africa

Several important events played a role in the development of higher education in francophone West Africa and the eventual establishment of the University of Dakar[4] in 1957, just before Senegal's independence: the establishment of the *Ecole de Médecine* (Medical School) in 1918, which is often cited by scholars as the "embryo" of higher education in the region (Bouche 1975; Ndiaye 2000); the Brazzaville Conference held in the Congo in 1944 (discussed in the following text); and the founding of the *Institut des Hautes Etudes de Dakar* (Institute of Higher Education, IHED) in 1950. These events reveal the *décalages* between colonial education policy and the reality on the ground; they also demonstrate how Senegalese worked within the constraints of the colonial framework to improvise solutions to advance the cause of higher education.

The medical school was founded by the French to train an African medical corps in 1918. Despite the colonial government's need for a "black medical faculty" articulated in official decrees as early as 1905 (Bouche 1975, 850), the school did not open until many years later. Debates between Paris and Dakar as to who would pay for the medical school accounted for some of the delay; however, the school only became a reality after Senegalese political activists and war veterans demanded that it be opened to compensate the more than 30,000 Senegalese who fought and died for France in World War I (Lunn 1999). By 1934, the medical school had graduated more than 400 students and was seen as a success on many levels.[5] However, by the 1930s, France began to shift resources away from the medical school and toward the creation of low-quality rural primary schools to support agricultural production and other exports in the run up to World War II (Coquery-Vidrovitch 1985; Sabatier 1978). This transition stalled progress in higher education.

The Brazzaville Conference, held in the Congo in 1944, gathered French colonial authorities who sought a new governance model that would sustain their colonies after World War II. It was seen by the French as a way to transition from "law and order" to "developmental" colonialism, indicating a new focus on the well-being of the colonized yet ruling out the possibility for their self-determination. The conference promoted training an indigenous workforce to lower the cost of labor and increase the profitability of the colonial operation through the expansion of primary and secondary schooling. Conference participants did not discuss higher education at Brazzaville, although they did briefly cite the medical school as it remained the only option for higher studies at that time.[6]

After a conference that laid no formal plans for the expansion of higher education, the threads of the discussion can be picked up in correspondence between colonial officials in Paris and Dakar that reference "The African University of Dakar."[7] Letters from March 1946 indicate the difficulty of a university project given its cost and commitments made at Brazzaville to prioritize primary and secondary education. These letters also indicate that colonial officials were under pressure to quell "local expectations," referring to the hopes harbored by Senegalese for a full-fledged university in Dakar.[8] But the number of African students who had gone to Paris in the 1930s and 1940s had reached a critical mass, and they began to advocate for expanded access to higher education in West Africa (Chafer 1997). Under pressure from these growing expectations and in keeping with the desire to promote a new, developmental colonialism, France organized classes in math and science for a cohort of 10 students in Dakar in 1948. By 1950, enrollments had grown to 94, and a small university institute was started by the colonial government.[9] Soon the IHED offered courses in law, general science, and liberal arts. In 1953, it added courses in medicine and pharmacy after merging with what was left of the medical school. Finally, in response to continued pressure from a newly formed Senegalese student union, the IHED was transformed into the University of Dakar in 1957.[10] At its opening, there were 581 students enrolled.

1960 to the CNES: From *Bricolage* to *Décalage* and the Reemergence of International Constraints in Senegal

With the transition to independence in 1960, Senegalese gained control of the government. Although they began to influence education policy more directly, the immediate post-colonial era remains a vivid illustration of the enduring structure of the French educational system and confirms the notion of *décalage* between the policy environment and the reality on the ground. There is no better example than the University of Dakar that, despite Senegal's independence, remained a French institution for many years. University leadership and faculty were French, as were most of its students and almost all of its funding. Only campus activism in Senegal in the wake of the 1968 worldwide student movement provoked a change. Senegalese students successfully advocated for an Africanization of the university. The percentage of Senegalese students at the university, typically 20 percent in the 1960s, increased to 50 percent of the student body in 1968; by the end of the decade, enrollment was 75 percent Senegalese. With respect to the faculty and staff, a succession of French university rectors finally ended when a Senegalese was appointed in 1971. By the late 1970s, Senegalese typically comprised 40 percent to 50 percent of the faculty and provided 70 percent of the university's budget (Seck 1994). The decade after 1968 and the gains made at the university during that time represent the success of *bricolage* as the Senegalese, in a relatively short period, transformed a French university into a national institution.

However, these gains would be challenged by Senegal's own education policy reforms. Two major attempts at school reform in Senegal, the 1971 *Loi d'orientation* (Orientation Law) and the 1981 *Etats Généraux de l'Education* (The General State of Education), illustrate that *décalage* was not limited to colonial education policy but extended into the post-colonial era. The 1971 law, written by a few people inside the Ministry of Education, was an attempt to address issues that were central to the student activism in Senegal in 1968. Likewise, the 1981 General State of Education drew its inspiration from a series of teachers' strikes in the late 1970s, which highlighted the declining quality of schools. The 1981 reform also criticized the 1971 law for its similarity to the colonial education system set up at Brazzaville. The 1981 meeting was a gathering of Senegalese from all walks of life, who came together over the period of a few years to craft a reform that would provide a clean break from the French educational system. The final document, 800 pages in length, advocated that education should be national and democratic and should function in the interest of people. Much like the colonial era, higher education was not the focus of either post-colonial education policy reform effort. However, the 1981 reform did acknowledge that the university budget could no longer support the growing student demand for higher education, and it called for a separate discussion on higher education. The 1981 reform was accepted by the government of Senegal with great fanfare and then promptly shelved when the government claimed it did not have the funding to implement such a wide reaching educational program.

Much like the colonial era, Senegalese took advantage of the *décalage* between the promised reform and inaction on the part of their government. At the university, government inaction, coupled with the worsening conditions on campus, led to years of student and faculty activism, which culminated in an *année blanche,* or cancelled school year in 1988. Although demands from the university community focused on the declining conditions for students and professors, they also began to dovetail with more widespread opposition to the IMF and World Bank and a general recognition of the reemergence of international influence over economic and social policy in Senegal. During this period, the discussion on higher education reform became linked to talks on structural adjustment. University reform, which had been an essentially Senegalese discussion in the 1980s, was increasingly driven by external donors, specifically the World Bank, who proposed the CNES in 1989.

The Limits of *Bricolage* in a New Era of Cooperation: Mapping the CNES Process and Understanding Its Outcome

By the early 1990s, the Senegalese had been advocating for progress in higher education for more than 70 years—from not only within the colonial structure but also with their own government when it failed to live up to higher education commitments. By this time, the situation at the university was approaching a critical breaking point. The institution, built for 3,500 students, was at many times its capacity with almost 25,000 students (Ndiaye 2003). The World Bank agreed with the government of Senegal on a funding package for a higher education reform process in hopes of improving conditions at the university. It seemed like good timing: the university needed help, the government backed the effort, and a large amount of funding would be associated with the reform, unlike the unimplemented 1981 reform where no funding was set aside. In addition, the participatory nature of the proposed consultation was meant to ensure that the reform agenda would be country-driven. However, after many years and millions of dollars spent, the reform failed. Given the economic and political power of the World Bank, it would be easy to conclude that it dominated higher education reform in Senegal, and there is ample evidence, discussed in the following text, to support this claim. However, the actual events of the CNES demonstrate something different, namely, that a well-educated and politically active local constituency became an effective counterbalance to inter/national forces attempting to dictate policy.

The CNES reform process illustrates how the dynamics of *décalage* and *bricolage* influence educational policy in the post-colonial era. In this case study, "local" is meant to describe the Senegalese university community—students, staff, faculty, university administration, and others—who participated in the CNES. The government of Senegal is examined as part of the "national" level, and "international" refers to forces outside of Senegal that influenced its higher education sector, such as France and the World Bank. To effectively map connections across these levels, surveys, interviews, and document analysis were used to plot the course of the reform and its many detours. Documents from the following sources were consulted: the Ministry of Education in Dakar; the World Bank in Dakar and Washington, DC; UCAD and national archives; personal archives of the CNES moderator and other participants; archives of the teachers union; student strike platforms; and newspapers from the early 1990s. In these archives, I examined official policy papers as well as meeting notes, mission reports, correspondence, annual reports, and strike platforms. Among the most valuable documents found were the sign-in sheets of those who participated in the CNES sessions. Surveys were then distributed to those who could be located, for a total of 54 former CNES participants. These participants, who were UCAD faculty, staff, government workers, and former students, were asked to evaluate (1) the success of the CNES; (2) which groups had the most influence over the reform; and (3) who benefited the most from the process. Follow-up interviews allowed participants to elaborate on their written responses about the roles they played in the CNES and the immediate and long-term outcome of the reform process. In all cases, the views of the participants and the reports in local newspapers were compared to official policy documents about the process. With respect to the long-term result of the CNES reform, interviews with CNES participants, as well as people in key positions at the university in 2006 (such as the Director of Finance and Budget, the Director of the Office of Research, the Director of Curriculum and Reform, and the Rector) helped to identify aspects of the reform that have survived this "failed" project.

The *Décalage* of Higher Education Reform in the Post-Colonial Era and the Limits of Participatory Reform

Although the CNES was often described by the World Bank as "country-driven" and part of a "national consultation," research reveals that it was neither. Although Senegalese higher education actors participated actively, the 23-point reform agenda did not reflect the recommendations of those actors. Instead, the agenda was principally an agreement between the World Bank and the government of Senegal. Thus, the implementation phase of the reform is a striking example of *décalage*. The improvement of teaching and research—the stated goal of the reform—was de-emphasized in favor of other "informal" objectives, which centered on cutting services to students, reducing the number of students at the university, and relaxing rules against private higher education in the country. With respect to being country-driven, documents and interviews suggest otherwise. They indicate that the World Bank not only proposed and financed the CNES reform, but it also suggested its structure and the core list of recommendations that would eventually be taken up by the government.

The publication and distribution of *Revitalisation de l'enseignement supérieur au Sénégal: Les enjeux de la réforme* (Revitalization of higher education in Senegal: The Stakes of reform), published by the World Bank in March 1992, caused a stir on the UCAD campus. Referred to by Senegalese as *"Les Enjeux"* ("The Stakes"), the document's impact reverberated throughout the UCAD campus and prompted the inauguration of the CNES process. The report was disturbing to many because it confirmed their suspicions that the World Bank's interest in higher education reform in Senegal was growing. When asked about how the CNES began, the vast majority of the UCAD faculty and staff cited this report as the origin of the CNES process, arguing that the World Bank was pushing higher education reform in Senegal. Further confirmation of this view was a meeting convened by the World Bank in Saly (a resort town in Senegal) in January 1992, just before the release of the report, which sought to identify strategies for restructuring higher education. This meeting was a follow-up to another held a year earlier between the Bank and the government of Senegal in which both sides sought solutions for the troubled economy.[11] Furthermore, the World Bank, in a 1996 report, took credit for the idea of the CNES and dated its interest in higher education reform in Senegal as far back as 1989 (World Bank 1996).

Once higher education reform was on the national agenda, the CNES was established by the president of Senegal in April 1992, who appointed Assane Seck, a former education minister and university professor, as moderator of the CNES. Although the World Bank provided US$100,000 to finance the CNES, it was organized and carried out by Senegalese with relatively little interference from the Bank for the first several months. During this time, Senegalese met in plenary sessions and working groups to determine recommendations for improving teaching and research. These groups continued to meet several times during a 14-month period, in a process that many described as a participatory series of meetings including more than 400 people from all walks of Senegalese life.

Although broad participation of government officials, university faculty and staff, teachers union representatives, and others can be confirmed via CNES sign-in sheets, a closer look at these lists reveals that a rather homogeneous core group of less than 100 people came together to carry out the actual work of the conference. Only 96 individuals participated in more than one plenary meeting during the 14 months of meetings. Furthermore, the survey administered to 54 of these participants reveals that faculty comprised more than 60 percent of these core participants. When broadened to include staff, the university community represented 80 percent of the CNES repeat attendees. In fact, one of the largest stakeholder groups, UCAD students, refused to participate in the CNES because they sensed that they were the real targets of the reforms in the eyes of the World Bank and the government of Senegal. In fact, only seven students from the *Université Gaston Berger*, a new university in Saint Louis in the north of Senegal, participated. In the end, the CNES was primarily a process in which some members of the UCAD faculty participated, rather than students, government representatives, parents, or others as originally intended, making the meetings not exactly the "national consultation" on higher education it claimed to be. Despite this uneven degree of participation, CNES participants generally described themselves as having had productive discussions guided by good will toward the reform process. They were aware of the World Bank's role, but they saw the CNES as an unprecedented opportunity to participate in a comprehensive discussion of higher education.

Although discussions had been free and open for many months, in the last plenary session of the CNES, it became evident that the influence on the final set of recommendations by the majority of participants—university-affiliated or not—was going to be limited. For instance, session participants were expecting to prioritize and consolidate the 90-plus recommendations they had been compiling over the course of 14 months. Instead, a team of educational experts funded by the World Bank worked separately to develop a 3-tome, 700-plus page report entitled *Propositions de réforme de l'université au Sénégal* (Proposals for university reform in Senegal) that culminated with 21 recommendations. Participants were aware that Bank consultants were involved with the process, but the report, distributed at the final CNES plenary, came as a surprise to most participants not only because of its size but also because they were expected to endorse its final recommendations rather than come up with their own. CNES participants debated until well into the morning hours to resolve the differences in the texts. Nevertheless, when recommendations were finally accepted by the government of Senegal, they were not those that had emanated from the CNES but rather—verbatim—the 21 recommendations set out in the Bank's report. The government added two more and, hence, the 23 reform measures that made up the CNES reform. CNES participants reported feeling frustrated that the recommendations appeared to be compiled with little regard for the CNES, and they were only able to place written caveats on the parts of these recommendations with which they did not agree. As one participant, a professor, described it, "The government got people to work on this for nothing. They also picked conclusions that went along with what they thought and with what the World Bank thought" (Interview, February 10, 2006). The publication of this report and the government's acceptance of it was the beginning of a series of *décalages* between reforms agreed upon during the CNES process and the implementation phase. These *décalages* would eventually result in a parallel reform process that had little to do with the CNES in its original form.

As the CNES moved into the implementation phase in the fall of 1994, the toughest and most drastic measures—which were, for the most part, aimed at students—became a reality, thus confirming students' suspicions about the CNES reform from the outset. Although both the Bank and the university stated that the reform would improve teaching and research, the first move in the implementation phase was not listed among the 23 recommendations. Instead, it was linked to what the Bank called an "informal objective" of the reform: a drastic reduction in the number of students at UCAD and an end to state scholarships and other student subsidies (World Bank 2003). This objective was hinted at in many of the documents related to the CNES, but it was clearly articulated in an *aide mémoire* (mission report). In this report the Bank stated, "[T]he University of Dakar has the objective to bring down enrollment to 15,000 students by the year 2000."[12] This meant cutting its enrollment by approximately 10,000 students. Also, the mission report made clear that funding for the CNES reforms would only be disbursed after significant progress was made in bringing down the number of students at the university and reducing the level of financial support to university students in general.

The desire for funding by the university administration and the government, and the prerequisite of reducing the UCAD student body, set up one of the most peculiar and little researched incidents in African higher education, the *année invalide* (invalid year). Following an intense three month period of student strikes and clashes with authorities, the university administration "invalidated" the 1993–1994 school year and evicted students from the campus and their dormitories by force. Students had only a few hours in which to gather their things and leave. Details of the *année invalide* and of what happened during this time period are neither included in official government documents nor are they mentioned in CNES documents. Newspaper articles and interviews with students themselves provide virtually the only accounts of what happened to students as a result of the CNES implementation. Fatigued by the strike and under pressure to begin the implementation of the reform with the *rentrée* (fall session) of 1994, UCAD Rector Souleymane Niang came up with the idea of the *année invalide*, counting the school year as a zero rather than just canceling it outright. The university had already lived through an *année blanche* (a cancelled school year) in 1988, and it was not prepared to pay the financial costs of redoing another school year. The "invalid" option ensured that the students who could not absorb another zero on their exams would be eliminated from the university forever rather than being allowed to repeat the exams as in the case of a cancelled school year. In interviews with former students, they estimated as many as 50 percent of their classmates ended their university careers as a result of the invalid year. These estimates are confirmed by other sources. A local Senegalese daily, *Le Soleil*, estimated in bold print and on the front page that 7,148 students did not return to UCAD for the *rentrée*.[13] A World Bank report (1996) put the number even higher, concluding that "the implementation of the new measures resulted in 9,007 students being expelled from UCAD at the beginning of 1994–1995" (13). Sources seem to concur that one-third to one-half of the students were forced to leave the university during this time. Even for those few students who were eventually able to re-enroll at UCAD, the toll was high. Mamadou Kebe, a student who made it through both the *année blanche* and the invalid year, explained, "To lose a year like that, in fact we are the generation that lost two years. The 1988 *année blanche* and the 1994 invalid year. Two years taken away from your career? It is like instead of being 35 you are 37" (Interview, February 24, 2006).

Other controversial reforms that were not part of the CNES agreement were aimed at faculty, such as raising the number of teaching hours, transforming sabbaticals into a competitive process instead of awarding them automatically, and a reorganization of the tenure process to eliminate assistant professors who did not finish their doctorates within a certain timeframe from the university. Unlike the students, though, the faculty, through their powerful union, negotiated with the government to delay implementation of these reforms and to apply them to newly hired faculty. Although there was some solidarity with the students on the part of faculty, professors generally went back to the classroom after their concerns were addressed by the government. The students, however, continued to protest the reform.

A more serious concern for many faculty members was the way CNES funding was to be controlled and distributed. As a result of the CNES process, funding was finally disbursed in 1997 via a new entity organized by the World Bank, the Higher Education Improvement Project (*Projet d'Amélioration de l'Enseignement Supérieur*, or PAES). The Bank appointed one person, the PAES administrator, to control the account. A Senegalese, he sat somewhere between the Ministry of Education and the Ministry of Finance in the Government of Senegal, but he was not part of the university structure. Although the PAES was conceived as the budget distribution arm of the CNES, it became a far larger operation, in essence a parallel reform effort. For example, under the PAES, additional reports and studies on higher education were commissioned, meetings were held, and final recommendations made to the government that differed significantly from the 23 recommendations linked to the CNES process. The most obvious difference was the PAES proposal for a new library. The PAES also purchased office and lab equipment and carried out faculty and staff training on campus. Although faculty would benefit from a new library and more equipment, they expressed concerns about the PAES because it was not part of the CNES process in which they had participated, and its funding targets were not at all linked to earlier discussions.

By the time the university received the first round of PAES funding, it had been three years since it had taken drastic measures in 1994 to expel students and cut scholarships and other student aid. In this period, student protests intensified, and in 1997 they made their first gains against the reform. The government, fatigued by the lag in World Bank funding and under pressure from students, agreed to reverse the more drastic measures of the CNES. The government restored some services and scholarships to students as well as loosened enrollment caps to accommodate a larger freshman class. Finally, in 2000, national elections all but ensured the end of the CNES. Senegal's new president, Abdoulaye Wade, was elected with significant student support, and, in exchange, he further increased student financial packages and called for open enrollment at UCAD. Today, the only remaining provisions of the original CNES reform in operation are reforms in budget disbursements to the university from the government and a limit on the number of times a student can fail exams and still remain at the university. With respect to initiatives that were part of the PAES, only the new library stands as a physical reminder of this attempt at participatory policy reform.

In 2003, the World Bank closed the project and published a final report suggesting that it was unsatisfactory and a failure. In the same report the government of Senegal called the project satisfactory. Surveys carried out in Senegal as part of this case study also revealed that most of the participants in the CNES saw it as somewhat of a success (56.5 percent). Only slightly more than 10 percent considered the CNES process a failure.

Follow-up interviews suggested that three things played into this perception of success for CNES participants: the new library, the change in budget dispersal, and the opportunity to come together and discuss the future of Senegalese higher education. On the question of who benefited from the reform, responses varied, with 39 percent of respondents indicating it was the government of Senegal, 37 percent indicating the students and 33 percent indicated the faculty. These figures get to the heart of problem. Although the reform was seen as a failure or somewhat successful depending on where one was situated, it is much more difficult to determine its tangible outcome and who benefited from the process. For example, the reform itself did not lead to many long-lasting benefits for UCAD. Faculty, staff, and students were able to prevent a wholesale restructuring of their university, through protests in the case of students and negotiations in the case of faculty. However, the participation of university actors did not influence the recommendations taken up by the government, and it was not through a "national consultation" but rather through protest and negotiation that participants in the reform influenced the outcome of the policy. While the university community was successful in disrupting this reform, their participation in the process has not brought about a viable alternative solution.

Implications and Conclusions: The Limits of *Bricolage* and the Power of *Décalage* in a Globalizing World

The 11 year (1992–2003), US$40 million CNES process and subsequent reforms achieved very little beyond temporarily shrinking class sizes, altering budget distribution processes, and building a new library. Although not insignificant changes, many observers would agree with the Bank that it was an "unsatisfactory" project. However, Ferguson (1994) argues that "even a 'failed' development project can bring about important structural changes" (275). He contends that what first appears to be an unintended result can become the central outcome of a development project. This notion was captured perfectly by a Senegalese professor of mathematics who, in assessing what the CNES reform achieved, said it "was a seminar in which minds were reformatted to think about higher education in another way" (Interview, January 16, 2006). The professor went on to suggest that things that were previously taken for granted in Senegalese society, such as US$10 university tuition, scholarships for every student, and state provision of the entire UCAD budget, were now being reconsidered by the university community. In addition, private higher education enrollment, one of the areas of interest for the World Bank in the CNES process, rose from 15 percent to 29 percent during the life of the CNES (World Bank 2003). Despite the failure of the reform and the strength of the university community to disrupt changes aimed at restructuring their university, the longer-term environment is not favorable to UCAD. The university has undoubtedly lost ground since the end of CNES, with a student body of nearly 60,000 and another threat in 2008 of an *année blanche*. With no solution to the enrollment problem at UCAD in sight and a continuation of the university's budget problems, the university is again in a difficult position as it tries to improvise a solution to these crises. The Bank, on the other hand, through a series of disjointed and failed projects that created enormous gaps between stated education policy and the reality on the ground, was able to change the higher education landscape in Senegal. The university is now considering privatization measures that were rejected years ago, and the institution is feeling some competition from a growing private higher education sector. Despite this, in 2008, Senegal opened two additional public universities and a regional university center (similar to a community college) to help with overflow at UCAD. They are again engaged in improvising a solution to their higher education crises and doing so by expanding the public higher education sector in defiance of World Bank economic and policy directives.

In the post-colonial dynamic, national governments, international institutions, and powerful local constituencies are all part of the policy equation, and interactions among these groups determine the success or failure of reform processes and other such programs. In the case of the CNES, its outcome was differentially influenced by the World Bank, the university community (primarily the students), and the government, which ultimately decided to reverse the reform. Masses of high school graduates continue to demand access to UCAD each year, as they have determined that higher education is the key to success in a globalizing world. Seen this way, the limits of *bricolage* are evident. Comprehensive solutions will be necessary if the public higher education system is to be an engine of national development for Senegal. One practical recommendation that could be drawn is that, given their role in the development of higher education during the colonial period as well as during CNES, students should be allowed to give input on higher education and their expectations for it.

The chapters in this section investigate the *bricolage* entailed in the formation and appropriation of policy, that is, the creative (and often contradictory) process by which different actors across multiple contexts negotiate the production of policy. Similar to Koyama (see chapter 1), I show how non-local actors such as the government of France and the World Bank are made material in local Senegalese interactions through object-actors such as colonial education policies or a list of 21 recommendations. Our chapters demonstrate the importance of examining in depth the vertical and horizontal social and political interactions through which policies are developed. Similar to Muro Phillips (see chapter 2), I demonstrate that the *décalage*, or significant gap, between stated educational policy and every-day actions reveals policy actors such as professors, students, and teachers who act in ways that reshape, hinder, or even derail policies.

Further, like the chapters in the third section of the volume, this chapter examines spaces created through attempts at participatory policy reform and the roles that local, national, and international actors and institutions have played within these spaces. This chapter concurs with Taylor and Wilkinson (see chapters 4 and 5) that access to information and the opportunity to participate are necessary but insufficient prerequisites for participation. This does not mean that participation is useless; as I have shown, it can have important, if unintended, consequences. What is needed are more spaces like the high school Hantzopoulos describes in chapter 6 in which faculty and administrators work collectively on problems that are jointly identified and addressed. The challenge of ensuring meaningful participation for all stakeholders is great, especially when institutions such as CNES depend heavily on the government and its international financial backers for support. Nevertheless, it is a challenge worth undertaking if participatory policy reform is to move from rhetoric to reality.

Notes

1. See, for example, the Partnership for Higher Education in Africa at http://www.foundation-partnership.org/.
2. For details, see the World Bank site on Tertiary Education at www.worldbank.org.
3. Ibid.
4. In the mid-1980s, the University of Dakar was renamed UCAD after the famous Senegalese Egyptologist Cheikh Anta Diop.
5. *L'Ecole de Médecine de L'Afrique Occidentale Française sa foundation (1918) á l'année (1934)* [The Medical School of French West Africa from its Founding in 1918 to 1934].
6. *Conférence Africaine Française de Brazzaville: Plan d'Enseignement* [French African Conference at Brazzaville: Teaching Plan], National Archives, Dakar, Series O 171 (31).
7. Letter from the High Commissioner of the Republic to the Minister of Overseas Territories, March 1946, Paris. National Archive, Dakar, O series 574 (31).

8. Letter from the Governor General of the AOF to the Minister of Overseas Territories, March 30, 1946, Dakar. National Archive, Dakar, O Series 574 (31).

9. Enrollment numbers found in a publication at IFAN entitled *La Revue Maratime*, 1959, 13.

10. *Association Générale des Etudiants de Dakar* (AGED) was formed in Dakar in 1950, the year the institute was inaugurated. Its primary concern was the quality of education at the institute, and its main goal was to transform the IHED into a full-fledged university.

11. *Seminaire sur la geston gouvernementale et le developpement economique*/Seminar on Government Management and Economic Development, February 7–8, 1991.

12. World Bank Mission report April 25 to May 17, 1995, 4. Located in the archive of the Ministry of Education in Dakar.

13. From Senegalese daily *Le Soleil* October 11, 1994, 1.

Chapter 3

AIDS and Edutainment
Inter/National Health Education in Tanzanian Secondary Schools

Tonya Muro

As the HIV/AIDS epidemic continues to ravage sub-Saharan Africa, numerous organizations have responded with a variety of approaches, platforms, and policy prescriptions meant to curb the spread of the virus. However, there has been little attention paid to the networks of actors who appropriate and localize the prescriptions from international organizations (see chapter 1). This chapter examines how local policy actors—namely an inter/national nongovernmental organization (NGO) and secondary school teachers—appropriated a media-based educational program adapted from an international health education effort. Specifically, I examine an inter/national organization, the *Femina-Health Information Project* (HIP), based in the United Republic of Tanzania, and its "hip" magazine, *Fema*, designed for secondary school-aged youth and reading clubs in their schools. Influenced by international donors and models, Femina HIP has endeavored to provide *edutainment* (education infused with popular cultural themes conveyed through mass media) in the form of magazines and television programs to inform Tanzanian youth about the causes and consequences of HIV/AIDS. However, Femina HIP has faced challenges in a number of secondary schools, where its materials are often censored before dissemination or not distributed at all because of local views regarding "appropriate" sex education.

This vertical case study investigates how inter/national health education materials are appropriated by local teachers whose views on sexuality and sex education for Tanzanian youth differ significantly from those of the staff—Tanzanian and non-Tanzanian—who develop health education magazines and television programs. Based on research at the Femina HIP headquarters in Dar es Salaam, the commercial capital of Tanzania, and in six secondary schools in the Kilimanjaro Region to the north, I explore the efficacy of HIV/AIDS edutainment produced at the inter/national level as it travels to distant schools where the youth lifestyle it "brands" conflicts with local sexual sensibilities. I argue that edutainment for HIV/AIDS prevention has paid inadequate attention to the significant sociocultural differences within and across nations regarding youth sexuality, thereby compromising the potential impact of youth-oriented materials such as Femina HIP's *Fema* magazine on its intended audience.

HIV/AIDS as Problem, Edutainment as Possibility

Across sub-Saharan Africa, a region devastated by the spread of HIV/AIDS, many international organizations have sought effective strategies for communicating health risks and safety measures to the general public. Since the late 1970s, many NGOs have adopted strategies of edutainment, which uses popular media to change dangerous attitudes and/or behaviors (Kaiser Family Foundation 2004). In the words of one sponsor of this approach, edutainment "involves incorporating an educational message into popular entertainment content in order to raise awareness, increase knowledge, create favorable attitudes, and ultimately motivate people to take socially responsible action in their own lives" (Kaiser Family Foundation 2004).

The risk of HIV/AIDS among Tanzanian youth has generated interest in edutainment at the national level. At present, the adult (ages 15–49) HIV prevalence rate stands at 6.2 percent, and there are approximately 1.4 million Tanzanians living with HIV (UNAIDS 2008). More than 60 percent of all new HIV infections are among young people aged 15–35, one of the most productive segments of society and one deemed likely to respond to edutainment programs (Tanzania Commission for AIDS 2008).

Edutainment serves as a striking contrast to traditional AIDS education in most Tanzanian schools (Vavrus 2003). Despite attempts by international organizations, such as UNESCO, to promote formal HIV/AIDS education in school settings, it remains inadequately taught for a number of reasons (UNESCO 2006). One problem is that the formal curriculum, which teachers are required to cover as fully as possible, includes little meaningful content on the subject of HIV/AIDS. Teachers admit that they lack sufficient time to teach about HIV/AIDS, and because it is not formally covered on the national exams that determine college entrance, they do not devote many lessons to it. Those lessons on HIV/AIDS that are presented are usually taught via rote learning because the most common approach to teaching in Tanzania is one "where teachers teach and students listen" (Stambach 1994, 368). Most HIV/AIDS instruction, such as instruction in other school subjects, involves students copying down definitions and diagrams in their notebooks without adequate time for discussion or elaboration. The HIV/AIDS instruction and general Sexual and Reproductive Health (SRH) education (also called Life Skills Education) that students do receive often comes through Peace Corps volunteers or local NGOs via special seminars and workshops (Health and HIV/AIDS Information for Volunteers, Peace Corps Web site; UNESCO 2006).[1] Therefore, the need for more HIV/AIDS education is evident; the form it should take, however, and how it should be delivered continues to be debated among all sectors of society.

Femina HIP: Inter/National Multimedia "Youth Lifestyle Brand"

Edutainment programs such as Femina HIP are an attempt to bridge the gap between youth-oriented AIDS pedagogy and standard Tanzanian classroom practice. Femina HIP was initiated in 1999, and it has been launched in more than 16,000 secondary schools across Tanzania. Its innovative approach uses youth-friendly peer drama, radio, television, and print media in both Swahili and English to entertain and educate young people about HIV/AIDS prevention and youth health and sexuality issues more broadly.

The edutainment materials that Femina HIP has produced are aimed at promoting a "youth lifestyle brand" similar to the "brand" found in some other African countries, most notably South Africa.[2] In the words of the executive director of Femina HIP, "What we represent in Tanzania is unique and revolutionary. We are creating platforms for youth to communicate throughout Tanzania for the first time about these very sensitive sexual issues. But, it goes beyond that. In essence, at Femina HIP we are creating a sustainable, positive 'lifestyle brand for youth'" (Fieldnotes, April 5, 2006). Though Femina HIP's original mission featured a Freirean-inspired orientation toward consciousness-raising (see, e.g., Bartlett 2005), thanks to a shift in donor priorities, Femina HIP's approach was revisited and recast during the months in 2006 that this fieldwork took place. The new, market-driven approach based on measurable outcomes sought to create a "lifestyle brand" for youth.

The desire to create a space for communication about sexuality among Tanzanian youth and a "positive lifestyle brand" associated with it exemplifies the influence of public health initiatives and donors both within and outside of Tanzania. Femina HIP, under the auspices of the East African Development Communication Foundation, began with a single donor in 1999, the Swedish International Development Cooperation Agency (SIDA), which has remained the primary funding stakeholder to date.[3] SIDA's HIV/AIDS funding strategy is called "Investing for Future Generations" and emphasizes the urgent need for a wide range of development actors to be involved in HIV/AIDS prevention (Tufte 2002; SIDA 2003). With that motto in mind, Femina HIP was conceived as a development initiative to combat HIV with input from experts in the media, health, and other sectors.

The mission and vision of Femina HIP was inspired by a variety of inter/national edutainment approaches (both within and outside of Tanzania), most notably UNICEF Tanzania and the South African Soul City Project.[4] Touted as one of the "leading health communication agents" worldwide, Soul City produces booklets and a reality soap opera series on sexual and reproductive health issues for young people (Fuglesang 2002, 141). Soul City also publishes a youth-oriented magazine that addresses such issues as HIV/AIDS, peer pressure, teen pregnancy and everything in between. *Fema,* clearly influenced by the Soul City project, is the most well-known form of AIDS education in the country because it comes in an easy-to-read format and addresses issues of central concern for youth.

Not only are Femina HIP's edutainment products influenced by non-Tanzanian models, but so, too, is its broader mission. It has several international funders, in addition to SIDA, many of whom are partners on other joint health initiatives in Tanzania, such as USAID and UNICEF. Although such funding allows for the widespread distribution of *Fema* and other Femina HIP materials, these various stakeholders influence the content of Femina HIP materials and do not always agree on their content. For example, USAID's policy during the period of study was to discourage, or even prohibit, the active promotion of condom use in favor of abstinence-only education.[5] However, other donor agencies (such as Family Health International who fund both Soul City and the *Fema* magazine) encourage comprehensive sexual education. Femina HIP uses different funding streams for the production of the magazine and TV show when the content may not be aligned with a donor's stance on some aspect of HIV/AIDS education (Fieldnotes, March 8, 2006). However, the donors who contribute to Femina HIP's operations agree that it is important to increase the production and distribution of print materials nationwide and expand the distribution of its television show.

Femina-Health Information Products: A Snapshot of the *Fema* Magazine

Ostensibly, the reach and scope of Femina HIP's products is widespread across Tanzania and reaches thousands of consumers due to the fact that it purports to serve the needs of local youth, despite perceived opposition from the older generation. Approximately 150,000 copies of Femina HIP's attractive, colorful, glossy *Fema* magazine are distributed quarterly to nearly 1,400 secondary schools in Tanzania every 3 months, and roughly 100,000 of the "*Si Mchezo!*" (No Joke!) magazines for less literate rural youth are provided to local NGOs, community centers, and health clinics on a bimonthly basis. Furthermore, booklets—adapted from the Soul City materials on various aspects of how HIV/AIDS affects the wider community—are published from time to time. These include a booklet entitled "Living with AIDS" and a popular version, in a cartoon format, of the national Tanzanian HIV/AIDS Policy (2001). The Femina TV talk show on health and well-being for adolescents is broadcast in thousands of homes, and a popular, interactive Web site known as *chezasalama.com* ("Play It Safe") receives more than 1,000 hits a week.

Fema began in 1999 at the inception of Femina HIP. The overall purpose of the 64-page quarterly magazine is to provide SRH information as well as to impart general life skills information, in both English and Swahili, to young people between the ages 15–30. The magazine features real-life stories of youth, its target audience, along with feature stories about Tanzanian celebrities. Each issue of the magazine centers on a theme or region of the country. Furthermore, the content is divided into various sections to mirror different aspects of the lives of Tanzanian youth in various parts of the country and how HIV/AIDS and Sexually Transmitted Infections (STIs) have affected them. Every issue includes the following sections: "Sexuality and Relationships," "Entertainment," "Lifestyles," "Body and Soul," "Work and Money," "Health Services," and "Living Positively with HIV." In addition, letters from readers and questions and answers on sexual and reproductive health appear in each edition. Coupled with this, regular crossword puzzles test SRH knowledge, and prizes can be won for the reader with the most correct answers. The model of *Fema* magazine was adapted from the Soul City magazines, which also instituted reading clubs all over South Africa.

Of particular interest in this study is a section in each magazine devoted to the extracurricular *Fema* Reading Clubs that describes how young people are using and adapting aspects of *Fema* in their schools. A *Fema* "User's Guide" has been published to accompany the magazine (but has not been widely distributed), offering lesson plans catering to teachers of biology and civics for activities and exercises for discussion and reflection on the content of the magazine. For example, one possible lesson plan combines basic numeracy and literacy skills by asking students to count different HIV prevention methods and write about them in their journals using previously published *Fema* articles. Although this is a welcome effort to support HIV/AIDS education in Tanzanian schools, the question is whether *Fema* is being used and, if so, how is it being appropriated by students and teachers in the classroom?

Research Methods and School Sites

This multisited, comparative case study utilized participant observation, interviews, oral and written questionnaires, and focus group discussions in two distinct research phases to study the development and utilization of materials produced by Femina HIP. I collected data at the Femina HIP offices in Dar es Salaam over a period of four months through participant observation, surveys, and interviews. I then traveled to six high schools in the Kilimanjaro Region for four months, where I used interviews, questionnaires, and focus group discussions to examine how teachers (and, in the broader study, students) understood and used the materials. In this way, both media producers' (Femina HIP) and media consumers' (teachers) perspectives on the edutainment approach and, specifically, *Fema*, were explored.

The teachers selected for the study taught at one of the six secondary schools in the Rural and Urban Moshi Districts in Kilimanjaro. Based on discussions with the Femina HIP director and other key personnel, these schools were identified as sites where students and teachers displayed a perceptible and active interest in Femina HIP materials. These were schools that had supposedly used *Fema* and other Femina HIP materials, established a Fema Reading Club, and taken several initiatives that indicated an open reception to Femina HIP's products. These six schools would, therefore, provide a fertile ground for exploring how teachers and students appropriated *Fema* in their schools.

Each of the schools was unique in some way. The sample included a community school (Moshi Rural, where school fees are collected from local communities), two government schools (Moshi Industrial and Moshi Boys), one nongovernmental school (Moshi Urban, financed by local economic institutions such as banks and businesses), one private school (Moshi Girls), and one religious school (Moshi Muslim Girls School). I included schools from both rural and urban districts to obtain a more complete picture of the entire Moshi district. The most salient differences among the schools had to do with the quality of the facilities, the gender breakdown of the students, the main religion of the students and teachers, the extent of the use of English or Swahili, and the teacher to student ratio. For instance, Moshi Rural School lacked many facilities, including a teachers' lounge; Moshi Muslim Girls had the best teacher to student ratio at 1:15 and the most modern amenities; and Moshi Boys had the highest percentage of students fluent in English.

Despite these differences, urban and rural secondary schools in Moshi have certain similarities common to most schools in Tanzania. The classrooms were comprised of wooden desks and chairs in rows, with one large blackboard in the front of the room and a single door and a few windows. The setting was fairly austere with no decorations on the walls except for the occasional chart, map, or photo of President Kikwete. The pedagogy displayed cast the teacher as the "sage on the stage" with a "banking" model of education where students are seen as rote repositories for knowledge delivery (Freire 1994; Stambach 2000; Vavrus 2003). Nevertheless, one school stood out from the rest in terms of facilities and access to Femina HIP materials: Moshi Girls. At this rural school, there was a well-functioning library where the girls were encouraged to read a variety of SRH resources, including the *Fema* magazine. All of the issues were on a table in the middle of the room, and the girls were allowed to pick up any copy to read, cover to cover (Fieldnotes, July 2, 2006). This was in striking contrast to the other schools, some of which did not have libraries and none of which had *Fema* on display or readily accessible for the students.

Contrasting Views of "Youth Lifestyles" at the Inter/national and Local Levels

Views from the Femina HIP Office

The Femina HIP staff is comprised of more than 22 young journalists, writers, TV and radio talk-show hosts, graphic designers, and public health educators from Canada, Kenya, Norway, South Africa, Tanzania, and the United Kingdom. This inter/national group of dedicated HIV/AIDS educators seeks to give young people the information and vocabulary with which to discuss sensitive health issues that have often been silenced by their families and in school. The staff was unanimous in their endorsement of edutainment as the ideal way to reach youth, as the TV talk-show host explained:

> Young people like entertainment, so using this method to communicate "extra large" issues like HIV and AIDS is a smart solution. It pins youth down to listen and ask questions, because it engages them at the emotional, individual level. At the end of the day, they have obtained knowledge that could save their lives if they choose to take it. (Interview, August 12, 2006)

The enthusiasm surrounding entertainment as a pedagogical—and life-saving—resource reflected the staff members' own experiences as media-savvy young professionals with a cosmopolitan view on HIV/AIDS education. The general view, though not shared by everyone, was that materials that worked well abroad, specifically in South Africa, would also be effective in Tanzania. Although some members of the Tanzanian Femina HIP staff recognized that there might be local opposition to some of the content in *Fema*, others disagreed.

To illustrate this sociocultural divide between South African and Tanzanian HIV/AIDS education policies and practices, I describe my observations from a *Fema* planning meeting with the editorial staff. The Femina HIP editorial team was discussing content from the Soul City booklets that they wanted to adapt for *Fema*, namely, material depicting condom use. Many of the staff members appeared wary of including articles and illustrations about condom use even though the Soul City team wrote widely about condoms in their materials, and past issues of *Fema* had included information about condoms in its descriptions of HIV/AIDS prevention resources. Several members of the editorial team suggested mentioning condom use indirectly in *Fema* articles with phrases like "use protection" and "safe sex practices." Yet two other staff members argued that Tanzania needed to "get with the times" because many young people were having sex with more than one partner. They further contended that the role of *Fema* was to bring these sensitive issues to the forefront because condom use was not openly discussed in the school environment; rather, it had been relegated to after-school presentations with local NGOs and this, they felt, was inadequate. They also pointed out that *Fema* had included some articles and cartoon illustrations about male and female condoms in the past, but they did concede that a number of secondary school administrators had objected to the content and refused to distribute the magazines to students.

I had also been told by Femina HIP staff that some teachers would rip out pages that referenced condoms and throw them away.

There was no apparent resolution to the matter, so the condom issue was tabled. After the meeting, I inquired about why the mention of condom use was taboo since it is integral to discussions of HIV/AIDS. One colleague commented that the majority of Tanzanians consider themselves to be religious (either Christian or Muslim), and therefore do not believe in sex outside of marriage. To bring condom use out into the open, she explained, would be the subject of extreme controversy and debate—and might encourage sex among youth. This, she argued, was in contrast to South Africa, which takes a much more "Western" view toward condom use and widely disseminates information about why and how condoms should be used to fight HIV/AIDS (Fieldnotes, May 20, 2006). Femina HIP's executive director also weighed in on this inter/national sociocultural predicament:

> While we believe it is our job to push boundaries and get people to open up [about sensitive topics like condoms], while we show the way, it is not in our interest, to, for instance, to get taken off the air [or to have our publications banned]. We need to be sensitive and understand where boundaries are for the future. (Interview, May 10, 2006)

The inter/national tension that the staff at Femina HIP headquarters faced between being direct about condom use and sexuality as one finds in some countries and being sensitive to local mores is further illustrated in the secondary schools in Kilimanjaro where secondary school teachers and students reinterpret *Fema*.

Views from Tanzanian Secondary Schools

I convened a small focus group of biology teachers at Moshi Industrial School to get their thoughts about the use of *Fema* magazine in their secondary school classrooms. The first teacher to arrive was the head of the biology department, who had been employed at the school for eight years. I asked her a series of questions, the first on her general impressions of the *Fema* magazine, and second on how the multimedia materials were being utilized in the classroom. The teacher responded by saying that she did not like the magazine very much and did not have time to use it during the school day with all of the other things she had to teach the students to prepare them for the national exams. I prodded a bit and asked her to elaborate on why she disliked *Fema*. She replied, "I think the magazine promotes sex amongst young people because of all of this talk about condoms. Our role as good Lutherans is to encourage our young people to go to church in order to stay well behaved." I then asked her if she taught HIV/AIDS in the classroom. She sighed and said, "I have mentioned AIDS a few times when we have talked about sexually transmitted diseases, but I have not been able to go into more detail than that. I assume that the students are well informed about HIV and AIDS already because of all of the NGO influence that has been in our schools in the past decade." Then I showed her a *Fema* Teacher's Guide containing concrete lesson plans on HIV/AIDS and sexuality in terms of general health. Her eyes perked up, and she concluded that there might be a place for *Fema* in her biology class, but that she would need formal training on how to effectively impart the information contained therein to her students without appearing to promote sex and condom use (Fieldnotes, August 9, 2006).

My research at six high schools in the Kilimanjaro region demonstrated that some teachers restrict the use of *Fema* or avoid it altogether because of their more conservative views regarding sexuality, which is in contrast to the overt mission and goals of the inter/national *Fema* magazine regarding sexual and reproductive health matters. Teachers at the majority of the schools were concerned about *Fema* content and appeared to censor it because they believed that talking about sexual and reproductive health might encourage young people to have sex. Therefore, students had limited access to viewing *Fema* content. A seasoned biology teacher from Moshi Industrial School reasoned:

[The magazine] is good, but some content causes sexual desires in students. Therefore, I don't allow my students to read the entire magazine; only sections of it. Don't insist much on the use of condoms because still those who use it sometimes they are infected [sic]. The most correct way especially for a student is to abstain from sex until they grow up because involvement in sex at their age distorts their learning and even their future life as expected mothers and fathers. They might also die. (Questionnaire, August 25, 2006)

Expressing a similar concern, one female biology teacher from Moshi Boys School cautioned:

It's a good magazine and TV show, but they should stop directing kids to use things which they have never used before. As a result, my students have never read an entire magazine. (Survey, August 10, 2006)

When asked for further clarification on what she meant by "use," she answered, "sex" (Interview, August 8, 2006). Many of the teachers I interviewed expressed similar concerns that the open discussions of sexuality in the magazine would encourage young people to experiment sexually.

Common beliefs about sex and religion among both Christians and Muslims in the Kilimanjaro region shaped teachers' attitudes toward teaching sex education in the classroom. Instead of using *Fema* to promote SRH, many of the teachers I interviewed focused on religion as a way to prevent HIV/AIDS. The following excerpts from focus group discussions and other data illustrate common views held by teachers:

> In terms of the ABCs of AIDS prevention, I tell my students that they should never have sexual intercourse before they are married, be faithful to their partners once they do get married, and always follow God's rules from the Holy Bible. There is no need to talk about anything else, since Jesus knows what is best for us. (Moshi Rural Focus Group Discussion, August 29, 2006)

> There is only one way to prevent HIV/AIDS—to receive Jesus Christ in your heart. Why is there a need to discuss anything else? (Survey, Moshi Boys female biology teacher, August 16, 2006)

> I teach my students to always abstain; be faithful and adhere to God's laws. (Moshi Industrial Focus Group Discussion, August 9, 2006)

> We must teach our youth about abstaining and controlling their desires rather than giving them different alternatives of fulfilling their desires. "When man fails to control his animal desires, he fails to differentiate good from bad." (from the Koran) (Interview, Moshi Muslim Girls school teacher, August 15, 2006)

These excerpts suggest that teachers felt a conflict between their own deeply held religious beliefs and teaching about AIDS prevention. As a result, abstinence and marital monogamy were emphasized and explicit SRH education was minimized or avoided altogether.

As indicated in the aforementioned editorial meeting at Femina HIP, condom use was a flash point for some teachers. In conversations with teachers about the "ABCs" (Abstinence, Be Faithful, and Condom Use) of AIDS prevention, the majority focused on the "A" and "B," but not on the "C." In fact, the majority of the educators with whom I interacted had objections to discussing condom use. For example, a female biology teacher from the Moshi Rural School implored:

> Please stop insisting that the use of condoms [is] the good way/method of stopping the transmission of HIV/AIDS in your publications. Some young people may want to practice sexual intercourse since there have been sections in the magazine about how to use condoms. I am hesitant to show

students the magazines for this reason, so I share only certain parts of *Fema*
with them. (Interview, August 29, 2006)

Teachers' religious objections to *Fema*'s content may well have retarded
the potential efficacy of the magazine as an edutainment endeavor.
However, there were other constraints at work as well. Specifically, because
HIV/AIDS did not figure prominently in the biology or civics curriculum,
teachers felt that spending time on the topic detracted from their efforts to
prepare students for high-stakes national exams. A biology teacher at the
Moshi Urban School reflected on the issue of curricular constraints:

> I have used [*Fema*] but not directly because only a few of the topics are
> related to what the syllabus has instructed me to teach. However, I have
> discussed these things with my students after classes in the classroom or
> during reading club time. (Focus Group Discussion, August 14, 2006)

Another teacher at the Moshi Urban School commented on the amount of
material to be taught in the biology and civics secondary school syllabi to
prepare students for their national exams. He, like other teachers, argued
that there is very little room to deviate from the prescribed curriculum
because of the examination (Focus Group Discussion, August 14, 2006).

As a result of such constraints, the use of *Fema* was often restricted to
after-school clubs in the schools I visited. Of the 22 teachers in the study, only
a handful in 5 out of the 6 schools used it in the classroom. In one school,
I found that the *Fema* materials were kept in a locked cabinet and only distributed when outside visitors (like myself) came to visit. In contrast to this
pattern, at Moshi Girls School, I found that the librarian was an enthusiast,
even though teachers at the school generally avoided the magazine. When I
first arrived, the librarian looked me squarely in the eye, grabbed my hand,
and told me that *Fema* was "saving lives" (Fieldnotes, August 2, 2006). This
teacher had lost several family members to AIDS. She stated,

> I want to try to help the next generation, namely my girl students. I want
> them to become more educated about HIV/AIDS before it's too late....
> [G]irls are often too shy to speak, so it will be my job as a *"bibi"* (Swahili
> word for grandmother) to encourage them to participate [in a *Fema* Reading
> Club]. Nevertheless, it's going to take a while to start the club because some
> of the parents don't like the magazine because of its explicit and open sexual
> content. (Interview, August 9, 2006)

The librarian further explained that the students asked both her and their
teachers serious questions about their sexual health because of the information provided in the magazine, and she felt that the teachers were learning

better ways of talking with their students about sexual and reproductive health matters as a result. The librarian was adamant that *Fema* was having a positive impact on teachers and students, as well as their family members (to a limited extent). She also provided details about how she uses the magazine with her young female students and the constraints that she has faced on when and how to use it:

> I use the *Fema* magazine with the young girls when they come to the library because they always have lots of questions about their bodies that they cannot ask anywhere else. They certainly don't feel comfortable speaking with their parents! I am very proud of the fact that we have all of the past issues of *Fema* around this table, so that the girls can pick up any copy that they want and read any article that they want. I am always here and available to answer their numerous questions and address their concerns, of which there are many. They just don't understand how their bodies work and all of the changes that they are going through. It really is such a pity that the *Fema* magazine is not distributed more widely or used in the classroom more often. Probably because the teachers lack training in how to exactly use them with their students. However, I am happy that our headmistress is open enough to allow me to talk to our girls at the library about very sensitive issues. (Interview, August 16, 2006)

This librarian's views indicate that Femina HIP materials are having an impact in some Tanzanian schools, but the reactions of the majority of teachers in this study suggest that there is a pronounced difference in perspectives on the appropriate content for *Fema* between the inter/national staff who design it and the educators who are supposed to use it in rural and urban schools.

Conclusion

This study has shown that the impact of inter/national AIDS edutainment resources on SRH education in schools is limited by the way teachers react to and understand the materials. The Kilimanjaro biology and civics teachers' attitudes about HIV/AIDS, youth sexuality, and *Fema* revolved around three principal issues: (1) limited teacher interest in using Femina HIP resources with students because they are unfamiliar with how to use them; (2) teacher ambivalence toward sexual and reproductive health education in the classroom because of their own religious views regarding youth sexuality; and (3) teacher skepticism about the utility of *Fema* and the importance of teaching about

HIV/AIDS because of mandates to adhere to curricular guidelines from the Tanzanian Ministry of Education to improve students' chances of passing the national exam. These issues impede the use of Femina HIP materials by teachers and students because teachers may not give their students access to them. The ability of materials such as *Fema* to promote sexual and reproductive health among secondary school students is, therefore, significantly compromised.

This research demonstrates that HIV/AIDS entertainment-education might have a place in secondary schools, but it is unlikely that this place will be biology or civics classrooms as the inter/national staff of Femina HIP assumed. The *Fema* Reading Clubs and other extracurricular methods of employing edutainment in schools might be one way of making better use of the magazine in a way that acknowledges teachers' multiple concerns about its use in the classroom. Furthermore, it might be useful for Femina HIP to recruit churches and mosques to get involved in sexual and reproductive health education, as many of the teachers who were interviewed for this research were local parishioners and seminarians. Moreover, teacher training workshops that not only show teachers how to use *Fema* but also engage Tanzanian teachers in the design and writing of its editorial content would help to bridge the gap between inter/national product designers and the organizations that fund them, and the teachers at the local level who appropriate these materials in ways that minimize their efficacy. Although this may mean altering the "youth lifestyle" that is being "branded" through edutainment products such as *Fema*, it is vital that HIV/AIDS educators attend to the multiple ways that health materials are interpreted by the actors who develop and who use them.

This chapter, like others in this volume, demonstrates the value of tracing how cultural artifacts such as policies (see chapters 1 and 8) or discourses (see chapters 4 and 5) get appropriated, recast, and redirected as they are translated over space and time. In this case, information seen by actors in Dar es Salaam (or Johannesburg or Stockholm) as liberating and health-promoting can easily be interpreted by actors in other contexts (like Kilimanjaro) as damaging or dangerous. Studying health education programs vertically and horizontally reveals the gap, or the *décalage* that Max explored in the previous chapter, between planned and enacted curriculum to reveal actors negotiating in ways that potentially contradict the original intent of the programs. Improving the efficacy of edutainment and other health education endeavors requires serious attention to the ways in which "consumers" are themselves productive in their responses to programming.

Notes

1. See the following Peace Corps Web site: http://www.peaccorps.gov/index. cfm?shell=learn.Whatvol.healthhiv.
2. See the following link for more information on lifestyle branding: http://www. hipsterrunoff.com/2008/05/what-is-lifestyle-brand.html.
3. Since 1999, the Femina-Health Information Project has received funding from more than 10 additional donor and technical support agencies, namely the Norwegian Agency for Development Cooperation, Development Corporation of Ireland, German Development Agency, the US Agency for International Development, Family Health International, Johns Hopkins University, UNAIDS, UN Development Program, UN Family Planning Agency/African Youth Association, Concern Tanzania, the Tanzanian Commission on AIDS, the Tanzanian Rapid Funding Envelope for AIDS, and the Foundation for Civil Society.
4. For more information about Soul City, see the following Web site: http://www. soulcity.org.za/.
5. US HIV/AIDS prevention policy for the past eight years has traditionally focused only on the A (abstinence) and B (be faithful), but not the C (condom use) in its guidelines. However, this is slowly changing. For more information, see: http://www.pepfarwatch.org/images/PEPFAR/globalaidsandhivpolicy.pdf.

Part 2

Exploring Participation in
Inter/National Development Discourse

Chapter 4

Questioning Participation
Exploring Discourses and Practices of Community Participation in Education Reform in Tanzania

Aleesha Taylor

> *There are two parallel systems working in Tanzania: life the way it is lived and the development game. The challenge is in figuring out how to connect the two . . . PEDP (Primary Education Development Programme), regardless of its faults, spells out what participation should look like. It is opening up space, and we need to grab that space.*
>
> *—(Rakesh Rajani, Fieldnotes, August 2, 2003)*

The Notion of Participation

During the past three decades, the notion of participation has become a pervasive component of education and international development discourses (Anderson 1998; Hickey and Mohan 2005). The term's proliferation in the policies and practices of development organizations might lead one to assume that decision-making power has shifted away from "experts" and policymakers to ordinary citizens. However, the view expressed by Rakesh Rajani, the former executive director of HakiElimu,[1] a Tanzanian civil society organization dedicated to education reform, cautions against such grand assumptions. His words capture the disjunction between development as a policymaking "game" and development as a lived reality. Yet rather than becoming disillusioned by the rhetoric and the reality of participation, as an activist, Rajani suggests that it is more useful to commandeer the spaces that the global discourse of participation has provided.

This chapter examines participation as a keyword in development discourse by asking *how* ordinary citizens participate in policy processes, such as the primary education reform in Tanzania known as the Primary Education Development Programme (PEDP). The concept of participation has achieved nearly universal, inherently positive status in development circles, yet it remains vaguely defined in Tanzania, as elsewhere. As with "inter- or multi-culturalism," "sustainability," and "democratization," explored elsewhere in this volume, participation has become a "floating signifier" lacking agreed on meaning: The term describes processes that are both constitutive of, and antithetical to, decision making by diverse groups of stakeholders (Anderson 1998). In the case of Tanzania, recent policies have tended to promote a vision of participation in which ordinary citizens enjoy increasing power in decision-making processes that determine how resources will be used and how social service delivery in local settings can be improved. Conversely, as this chapter shows, descriptions from ordinary citizens often paint a picture of participation as something more mundane and prescribed, such as being involved in construction activities or in making financial contributions to schools.

HakiElimu, whose education programs will be examined in the following pages, seeks to enable ordinary citizens to transform schools through participation in policymaking, create a national movement toward social and educational change, and influence educational policy and practice. To these ends, HakiElimu works locally, regionally, and nationally to deepen participation and transparency in educational policymaking in the following ways: fostering a national dialogue around education reform; informing citizens about their rights and responsibilities to involve themselves in decision-making processes; and building capacities of local governments, local associations, and school committees to take advantage of the space for involvement created through PEDP.

HakiElimu's efforts to "grab the space" created through existing policy frameworks to build the capacity and effectiveness of school committees demonstrate the contradictions experienced by international development actors, national policymakers, and local practitioners of participation in policy and practice. Through this vertical case study of primary education reform in Tanzania, I argue that the discourse of participation allows for a carefully delimited involvement of local actors while not permitting their priorities to influence policy decisions.

Historical Context: Education Reform in Tanzania

The PEDP formed the basis of Tanzania's primary education reform efforts from 2002 through 2007. PEDP served as the initial education framework to nationalize and realize the global Education for All (EFA) agenda in Tanzania. It was also the education component of the national Poverty Reduction Strategy, which was designed to eliminate a portion of the country's external debt if the savings were put toward specific sectors, such as primary education. In addition to increasing access to education, PEDP also sought to increase the level of citizen engagement in schools and education reform. Like other recent national policies, it highlights the importance of, and outlines the mechanisms for, meaningful involvement of local populations in the process of educational decision making. Specifically, PEDP identified democratic school committees as the vehicle for effective community participation in education governance (World Bank 2001). PEDP represented the contemporary inter/national climate of education reform through decentralization and increased involvement of local populations in governance processes. PEDP reflected EFA goals to improve educational quality, increase school retention, and expand educational access at the primary level across the country. Reflecting the overarching thrust of the EFA agenda, the primary focus of PEDP was increasing access to primary school in Tanzania; therefore, it sought to remove all fees and levies that restrict educational access. Improved quality of educational delivery was sought through increased community participation, and democratic functioning of school committees was identified as a crucial means of increasing involvement of community stakeholders in educational governance (World Bank 2001).

Through PEDP, the government of Tanzania expanded access to education by increasing classroom construction and instituting the abolition of school fees. Education quality was meant to be enhanced through a US$10 per pupil grant for each primary school. PEDP sought to increase community participation by encouraging further decentralization and devolution of power to the district and school levels. To this end, the policy called for the direct allocation of funds for primary education to the bank accounts of individual primary schools. Furthermore, PEDP attempted to increase the effectiveness of school committees, which played a pivotal role in the reform strategy. School committees had the responsibility to oversee the proper use of funds allocated to school accounts, disseminate information to the wider community, determine priorities, and manage program implementation. To operate successfully, school committees had to first be reorganized to become democratic and representative of all members of the school community, including teachers, parents, administrators, pupils, village representatives, and marginalized groups, such as the disabled and poorer members of the community (World Bank 2001).

Democratic school committees were identified in PEDP as the mechanism for creating meaningful local involvement in educational governance and the vehicle for reform, local ownership, and local involvement in decision making (United Republic of Tanzania [URT] 2001; World Bank 2001). Their identification as a principal means of community participation is an aspect of the "space" that HakiElimu's former executive director asserted could be "grabbed." The organization's emphasis on promoting the democratization and proper functioning of school committees is one such attempt to grab this space and enable the meaningful involvement of ordinary citizens in governance processes. Its efforts in this area sought to equip school-level actors with the skills and information necessary to participate in educational governance and decision making.

The processes of engaging school committees in education governance were situated within Tanzania's national decentralization framework, the Local Government Reform Program (LGRP), which restructured systems for local governance that are responsible for social service delivery (Makongo and Mbilinyi 2003). The LGRP is outlined in the Policy Paper on Local Government Reform (URT 1998), a national policy that identifies decentralization by devolution as the governance mechanism for broadening participation in development and the delivery of social services in Tanzania. The overarching goal of local government reform in Tanzania continues to be to facilitate the government's efforts to reduce poverty, and this policy, along with PEDP, is couched within the national Poverty Reduction Strategy. These policies and processes (i.e., decentralization, participation, and poverty reduction) reflect donor priorities in international development and also constitute preconditions for donor aid from certain multilateral and bilateral organizations (Mosse 2005; Vavrus 2005).

In 2002, HakiElimu initiated a program in the Ukerewe District, a cluster of rural islands in Lake Victoria in the Mwanza region of western Tanzania. The program attempted to enable what HakiElimu described as "meaningful participation," or increased participation by ordinary citizens in governance and decision making. This intervention presented a unique opportunity to understand *what actually happens* when people participate, to gain access to the experiences of ordinary citizens as they participate in governance processes, and to highlight the conflicts and issues that are often overlooked by traditional policy studies of social interventions.

Local government and education reforms have worked together to broaden opportunities for democratic engagement for citizens in Ukerewe. Together, the reforms have sought to democratize governance at the district, village, and school levels. At the district level, district council employees have been trained to enable them to do their jobs more efficiently and incorporate more citizens in decision-making processes. To this end, the locus of control for district development planning—which consists of consolidating plans for each of Ukerewe's 74 villages into a comprehensive plan or funding proposal—has been decentralized from regional to district authorities. In villages across the Ukerewe islands, recent reforms have ostensibly resulted in village governments that are democratically elected and accountable to the communities they serve. Citizens have gotten opportunities to identify their needs and democratically plan how their resources should be utilized. At the school level, school committees have been democratized to ensure that they are comprised of community representatives and trained to oversee all aspects of school functioning—including procurement and construction activities. They are now meant to be accountable to village governments and solely responsible for all financial decisions related to the school. Before local government reform, citizens were excluded from decision-making processes and were not afforded access to information regarding resources. The LGRP, therefore, worked together with PEDP to enable community participation in education governance by outlining the statutory governance through which citizens should be engaged and decision making decentralized. The remainder of this chapter contrasts these various national interpretations of participation to those at the international and local levels.

The Paradox of Participation

Participation has become a keyword within international educational development discourse through its institutionalization by governmental and nongovernmental organizations (NGOs). Yet, as discourse analysts have shown, discourses generally obscure as much as they reveal. Further, they often enhance the power of one social group or institution over another (Lemke 1995; Smith 2005). Development discourse itself is a powerful regulator of international educational policies and politics, framing some issues as "problems" while sidelining others, and specifying "solutions" through a process of naming and classifying certain outcomes as, a priori, more desirable than others (Escobar 1995; Robinson-Pant 2001; Smith 2005). Development discourse, always itself evolving, influences what is nameable and hence actionable, and in this way it directly influences practice.

The increasing prevalence of participation in development discourse generally and in international educational policy documents specifically must be noted as an important characteristic of current thinking in public policy and international affairs. Green (2003) has shown that contemporary development projects and policies in diverse settings are strikingly similar, "despite the introduction of participatory planning methodologies which are intended to constitute the mechanism through which beneficiaries can become involved in the design and implementations of interventions which affect them" (123). This chapter considers such a paradox in an educational development reform grounded in participatory discourse and practice that nonetheless (de)legitimated actors and actions in fairly scripted ways.

Highlighting the discursive nature of development reveals the power dynamics inherent in international development policy when countries dependent on foreign aid must respond to the priorities of international donors while also considering domestic constituencies, whose development priorities often differ. Attention to "policy discourse" requires analysts to juxtapose policymakers and their conceptual frameworks to those of various citizen groups, who not only experience them in their daily lives but also reinterpret them. This analysis results in multiple meanings and complex understandings of such concepts as decentralization and participation, as the following sections reveal (Torgerson 2003). Highlighting international policies and processes obliges us, from a vertical case study approach, to also examine their on-the-ground interpretations (Eastwood 2006).

Methods and Methodology

The vertical case study approach provided a framework that allowed me to consider different positions and perspectives and observe the conflicts and contestations that occurred within a participatory education project. In particular, it helped me to analyze international discourses as enacted at the national and local levels. Being an international researcher working with a national NGO to study a local education reform activities sponsored and shaped by international development organizations helped me to illuminate the power struggles of policy actors within and among these different levels. The vertical case study approach provided a means for me to explore what Cooper and Packard (1997) describe as the "subtle interplay of national policy, foundations with the financial resources to shape intellectual inquiry, and the operations of programs in the field" (29). In this case, the "field" was constituted by HakiElimu headquarters in Dar es Salaam and by its education project in Ukerewe.

My examination of participation explored the following questions: (1) How is the concept of community participation constructed and contested by different groups of local, national, and international policy actors engaged in education reform in Tanzania? (2) How are these meanings expressed in practice at the local level? Through data collected primarily through participant observation, I addressed these questions by situating the discourse and practice of participation within a broader inter/national context, while simultaneously studying the implementation of policies and processes that sought to enable participation at the local level. I focused on HakiElimu, a key national institution in the Tanzanian education reform process, while also studying the local and international stakeholders who are influenced by and have influenced education policymaking in Tanzania and elsewhere in the global South. I spent nine months (November 2004 to August 2005) working with HakiElimu in their headquarters and their field office in Ukerewe; in addition, I made a one-week follow-up visit to Ukerewe in October 2005.

Participant observation in Ukerewe enabled access to the everyday worlds of citizens, who, because of recent reforms, reportedly have more opportunities to take part in the decision-making processes that govern them and their resources. A focus on the experiences of ordinary citizens provided a basis from which to explore and trace how everyday life may be "oriented to relevancies beyond a particular setting" (Campbell and Manicom 1995, 8). Local experience is, therefore, the basis from which to explore the various ways people's lives become organized by discourses and policies that originate far beyond their localities and national boundaries (Campbell and Gregor 2002). In other words, my interest in Ukerewe was not limited to residents' descriptions of what participation should be or how it was carried out by them. I was also interested in noting how everyday citizens' experiences with participation were shaped by, and relevant or accountable to, powers far beyond Ukerewe, including HakiElimu, the Tanzanian Ministry of Education, and the World Bank.

In addition to participant observation, I conducted 39 semi-structured interviews with educational stakeholders, including donor representatives, government officials (from national, regional, district, and subdistrict levels), national and district-level civil society actors, university professors, and K-12 school leaders. Six additional in-depth interviews were carried out with selected key informants chosen for the perspective, extent, and history of their participation in education governance. The two females and four males included members of school committees and village governments, a retired teacher, district and subdistrict officials, and a Community Information Volunteer (CIV).[2]

To complement participant observation and interviewing, I also carried out six semi-formal focus group discussions of four to twelve participants in Ukerewe with school committees, village government representatives, and a community group of key informants. Moreover, I employed document analysis to explore the ways in which participation was presented in policy documents and the means through which it has been sought.[3] Documents reviewed included district-level reports and development plans, national policies and annual reports (related to primary education, local government reform, rural development, and poverty reduction), and international framing documents related to EFA and decentralization. Together, these methods allowed me to explore the network of policy actors, including the policies themselves, that have shaped the meaning of participation and to compare how these actors envision the practice of participation at the local level.

Questioning Participation

Access To Information: Does Informing Lead to Decision Making?

Armed with information about their expanded roles in governance, as well as reports on school and village finances, some citizens in Ukerewe have been able to play a greater role in village and school governance processes and hold their leaders accountable. Information is, indeed, power: Without knowledge of policies and procedures, governance remains obscured, and citizens remain uninformed of their rights and the responsibilities their leaders are supposed to uphold. A member of Ukerewe's district council explained the importance of information in this way:

> Since local government reform, we are now able to do our work better because we have been informed about the laws and proper procedures. If you are illiterate on how to run the government, you cannot do your job properly. If the DED [District Executive Director] makes a mistake you would never know. Now we know. So we can do our jobs properly. (Interview, February 24, 2005)

The critical role of information in enabling meaningful local participation is also reflected in official definitions of democratic participation, as seen in documents outlining local government reform and the institutional arrangements of PEDP. At the village level, school committees have the responsibility to "effectively communicate educational information to all parents, pupils, community stakeholders, and to the village, ward/mtaa [neighborhood], and LGAs [local government authorities]" (URT 2001, 16). Ward level actors are responsible to "share information with, and facilitate the participation of all parents and the wider community in realizing the PEDP objectives" (17). District authorities have the responsibility to "effectively communicate educational information to village, wards, schools and other local stakeholder groups, as well as to regional and national levels" (18). PEDP contains similar statements regarding the importance of information for regional, national, and civil society actors as well.

Despite this consensus on the importance of information, observations and discussions with research participants in Ukerewe raise questions about the role of information alone to alter existing power relations in governance processes. It is tempting to believe that because citizens have greater access to information, including accessible descriptions of national policies and financial reports for their districts, villages, and schools, they now have increased power to control their resources. Although it is clear that simply increasing access to information does not automatically transform development and place it in the hands of ordinary citizens, it is imperative to explore what people have been able or inspired to do with this information to effect change in educational decision making in their communities.

In the context of Tanzania's post-colonial development, the purpose and role of information about development projects has varied over time. Though echoed by various participants, this reality was best illustrated by a research participant—a district official—when specifically asked to describe participation today compared to 20 years ago. He cautioned me about the intricacies of the concept, which he illustrated by using two related Swahili terms. The participant explained that in English, one term is used for participation. However, in Swahili, participation is related to two terms with two different meanings: *ushiriki* and *ushirikishwaji*. "There is '*ushirikishwaji*'—for people from above to involve those below by giving orders and asking them to implement them. Then there is '*ushiriki*'–to involve oneself, to be able to plan and make decisions" (Interview, May 5, 2005). He explained that, historically, participation in Tanzania is best characterized by *ushirikishwaji* because the impetus for development projects was often outside of village structures, and decisions about social service projects were made with little input from recipient communities. This was particularly true during the immediate post-colonial period, at which time the central government prioritized social service delivery and believed that a highly centralized system was the best way to provide services and enable national development.

This distinction between *ushiriki* and *ushirikishwaji* is useful in teasing out the role of information in participation and considering what other tools may be employed to "involve citizens from above." In the former characterization of participation, wherein people are able to plan and make decisions, information would have a facilitative role enabling more informed and effective decision making. The latter term, *ushirikishwaji*, reflects the way in which participation took place before recent reforms. Information had a more coercive role in that information was presented after decisions had been made by centralized authorities. Rather than ensuring that citizens were aware of village finances or the responsibilities of their leaders, for instance, information largely consisted of directives from above and instructions on carrying out decisions made elsewhere by others.

In Tanzania's contemporary policy context, information is meant to facilitate transparency and accountability. The participants in this study, including district and village officials as well as parents who held no official position in schools, described education and local government reforms and the work of organizations such as HakiElimu as disrupting the status quo as recently as five years ago. Still, it is useful to apply the distinction of *ushiriki* versus *ushirikishwaji* to contemporary policy processes that purport to facilitate participation in decision making. The next section specifically focuses on recent reforms that have emphasized broadened access to governance processes through participation in meetings and collaborative planning mechanisms.

Does Participation in Governance Equate to Participation in Decision Making?

The LGRP has been credited with enabling local governments to adopt and adhere to good governance principles. The increased frequency of community participation in statutory meetings is held up as evidence of democratic local governance across Tanzania.[4] These meetings are not only taking place with greater frequency, but citizens also contribute ideas on how community resources should be utilized and speak more freely than they did in the past. Participants in Ukerewe explained that ordinary citizens are now able to decide what can and should be done in their villages and then inform those at higher levels of their preferences and priorities. Processes directly related to LGRP have been credited with this shift. A district representative noted:

> After local government reform, each village makes its own plan. They create their own budget. Before, plans were made at district without input

from villages. It was like a district project, not a village project, because they were not involved. Now each village sits to discuss projects. (Interview, February 15, 2005)

Intrigued by the prospect of such encouraging changes taking place in a relatively short period, I typically followed descriptions of changes in community participation with a question about the types of decisions that ordinary citizens were now able to make. In terms of education, participants explained that village assemblies provided space for citizens to discuss the number of classrooms and teachers' houses to be requested in school development plans, to decide on the nature and amount of contributions for classroom construction, and to determine whether to build secondary schools. A village government representative described the unprecedented number of community meetings held in the previous year, "Last year there were more than four village assemblies because they were building classrooms for the primary school and starting the secondary school" (Focus Group Discussion, February 23, 2005).

The extent of citizen agency within such processes related to local government reform can be distinguished through consideration of *ushiriki* and *ushirikishwaji*. Participatory processes are used to create school development plans, which are later compiled into village development plans. Descriptions of what participation should look like, which are consistent with the global push toward good governance and decentralization by devolution of power, paint a picture more in line with *ushiriki*, in which ordinary citizens involve themselves in governance processes with the goal of impacting decisions. However, if we consider that these participatory processes are now required by donor agencies or central governments as a prerequisite for funding, the question of agency becomes more complex because it appears that Tanzanians are once again "being involved from above." An interview with a retired professor from the University of Dar es Salaam with extensive experience in educational policy articulated these concerns:

My major concern is that when they [government and donors] talk about participation now, it's participation after the policy has been made. For example, in PEDP the kind of participation the document talks about is mobilizing communities to build classrooms, rather than participation towards formulation towards the policy itself. This is also true of the PRSP and PEDP as well. If you ask anyone what PRSP is, they don't know because they were never involved in discussing it. So, it is policies made for them by us, the elite class sitting in Dar. A condition in PEDP is that schools can only get funds if they have a school development plan, and it is said that there must be community involvement in developing these plans. And if you have been in Tanzania for a while and look at the situation, you would see that participatory planning is not feasible. It's just not happening. Imagine that you are a District Education Officer. If you get 100 plans, how do you sit and make a district education plan based on all these individual school plans? And we are assuming that you will get all the plans in time. You just will not have the resources to do it. Then, on what basis do you make a decision on which school gets what allocation? So the planning and decisions about resources are two completely separate processes. (Interview, May 23, 2005)

This interview illustrates the similarities between international and national development processes in Tanzania. According to the institutional arrangements outlined in PEDP, which is itself a national policy influenced by international debt relief and educational priorities, local school development plans provide "the basis for all decisions regarding improvements in the quality and delivery of education, and new construction in the school" (URT 2001, 13). School committees are responsible for creating the plans based on a three-year cycle that coincides with that of village development plans. Once the plans are drafted, they are forwarded to the village council to be discussed in village assembly, where all community members are able to give suggestions and are consolidated by the District Education Officer. Thus, the policy suggests that resource allocations to primary schools are based upon democratically created school development plans, and participants did, indeed, describe the processes through which they contributed to the creation of these plans.

The inter/national embrace of this form of participation by local community groups indicates a degree of alignment among policy actors at all levels; however, when asked whether they thought their plans had any bearing on resource allocation, school committee members often expressed doubt. During one focus group discussion with the school committee for a primary school in Nansio, Ukerewe's district center, I noted that participants said the classrooms and teachers' houses that had been allocated by the district were far fewer than what had been requested by the school committee. One committee member explained that while they requested 36 classrooms, a library, and 5 offices, they are hoping to receive 2 classrooms (Focus Group Discussion, February 24, 2005).

Similar data from other participants implies that participation in Tanzania, although couched in a rhetoric of participation, continues to resemble *ushirikishwaji*, or involvement from above. Instead of information being used as the tool to involve ordinary citizens in development, for example by helping to formulate policy mandates and directives, citizens are involved through processes that appear to present them with decision-making power yet may essentially have little effect on actual decisions. This apparent disconnect was made clear during a visit to the district planning officer, which also exemplified the distinction between divergent processes for planning and decision making described earlier by the retired professor. The district planning officer went into great detail in describing the village and school planning processes, and he pointed to several stacks of plans that sat in the corner. When I inquired about the process of allocating the number of classrooms to be constructed in any given village, he explained that he is able to make recommendations on this, given his knowledge of the district and their needs. As the professor predicted, this district official did not actually read the large number of plans submitted to him but rather made decisions based on his "knowledge" of the local situation.

Another example of questionable participation is evident in the processes that precede classroom construction by community-level stakeholders. According to the PEDP initiative, the government is responsible for "topping up" community contributions for the construction of teachers' houses and classrooms. This topping up refers to the provision of materials not locally available, such as iron sheets and nails. Therefore, community members are responsible for contributing materials that are easier to obtain, such as sand, stones, and water. Community participation, in this sense, essentially takes the form of indirect school fees that add to the private costs of education for parents, who must make these material contributions. When asked to describe the decisions that communities are able to make regarding construction activities, members of school committees and village governments gave examples of how communities "decided" to contribute such resources. For example, a parent and school committee member explained:

> We have good community participation in our school. We are building three classrooms. We have more than 100 children in some classes. The classrooms are in disrepair. We talk about the school during village assemblies. The community decided how many classrooms we should build. We decided that we would provide some of the materials ourselves. Women are bringing water for the classrooms. The men bring sand and stones. (Interview, February 17, 2005)

Despite this involvement of the community, many "decisions" noted by participants actually reflected government directives about what ordinary citizens must contribute to construction activities. For example, recent policy reforms regarding local governance processes require that communities discuss and agree on any development initiative that will take place in their villages. Therefore, contributions of resources and the means for obtaining them must be authorized during village assemblies. This suggests that the discursive democratization of local governance processes masks the distinction between local control over decision making independent of government wishes, and community consensus out of necessity to conform to government policy to get needed resources for local development projects. In other words, what was assumed to be participation in decision making essentially entailed communities conforming to policy dictates from above. Although participants generally assumed they were exerting their power and control over their resources through school committees or village assemblies, they were, in fact, ratifying government directives. The views of local participants about their influence over decisions relating to resource allocation lead one to question inter/national discourses and practices surrounding participation because of the striking similarity of stated local priorities and national policy directives as well as the reservations of local participants about the impact of school and village development plans on policy decisions. It appears that participatory practices around education reform reflect *ushirikishwaji* rather than *ushiriki*. The next section looks at the nature of school committee functioning and raises further questions about the nature of participation in education reform.

Constrained School Committee Decision Making: Buffet or Prix Fix?

School committees have been identified in both the PEDP and the LGRP as the locus of decision making for primary schools in Tanzania. A cursory glance at school committee democratization and capacity building in Ukerewe suggests that they have been enabled to fill this crucial decision-making role because of their democratic make-up, the training they have received to educate them on their roles and responsibilities, and their stated importance within the primary education and local government reform processes. School committees in Tanzania have been given enormous decision-making responsibilities compared to ten years ago, when decisions in primary schools were made by one or two individuals rather than through an openly democratic process. The responsibilities of these new committees, as outlined in the PEDP document, include sensitizing and involving all pupils, parents, and school staff with respect to the roles they can play in maximizing the benefits of primary school; working closely with the head teacher and other teachers to prepare a Whole School Development Plan; facilitating planning, budgeting, and implementation of the PEDP-funded activities; and, effectively communicating educational information to all parents, pupils, and community stakeholders (URT 2001).

Although school committees remain accountable to the village government, they are responsible for all financial decisions in primary schools, which is a very important change from the recent past. Yet one should be cautious about accepting at face value the impression that school committees were accorded full control of how PEDP funds were used at their respective schools. Instead, it is important to examine the nature of their decision-making power: simply put, were school committees at a buffet where they could choose to purchase anything they wished, or were they selecting educational entrees from a limited prix fix menu?

The regulations related to the way PEDP funds were supposed to be utilized at the school level provide insight into this question. There were three separate tracks of PEDP funding to schools: a capacity building fund, development grants, and capitation grants to fund teaching and learning materials. Capacity building funds were provided for school committee training. Development grants, also referred to as investment grants, were earmarked for construction of classrooms, latrines, teachers' houses, water tanks, desks, and other furniture. Capitation grants, which were supposed to total US$10 per pupil per annum, were intended to improve education quality and cover renovations, teaching aids, administrative materials, and examinations. The PEDP text stated,

> In order to introduce a reliable income stream for essential non-salary expenses at school level, a Capitation Grant equivalent to US$10 per enrolled child will be instituted nationwide as of January 2002. Of this, US$4 will initially be sent to the district to enable schools to acquire textbooks and other teaching and learning materials. The remaining [US]$6 will be disbursed to schools through the district council, and school committees will decide how best to use the funds. (URT 2001, 11)

Although school committees were supposed to receive US$6 per pupil to distribute as they wish, participants noted that the government had already decided how the funds should be allocated. For instance, in a focus group with a school committee near the district center, an assistant head teacher echoed a common sentiment when he mentioned that their PEDP funds for books had not been sufficient. I asked the group why other monies could not be utilized to make up for the shortage, and the same participant responded:

> Well, the money is not enough in general. PEDP has helped us very much. There are more desks and supplies. We have two new classrooms. Students have to share books, but there are more [students] than there used to be. Teachers have chalk now. Chalk used to be a very big problem. We have more supplies. This is true. There is still a shortage. They [district officials] tell us how to use the money that we get. We can't say that we will focus on books this year or classrooms next year. They are very clear in telling us

how much money can be used for various things, like construction, paper, books, and things like this. (Focus Group Discussion, February 24, 2006)

The PEDP document does, in fact, restrict the decision-making authority of school governing bodies by outlining how the capitation grants should be allocated. Two dollars was reserved for facility repairs; four dollars for learning (reading) materials; two dollars for chalk, exercise books, pens, and pencils; one dollar for administrative materials; and one dollar for examination paper (URT 2001).

Experts on Tanzanian education reform who participated in this study explained that this type of a priori identification of how resources should be spent is an aspect of a highly centralized governance system. They argued that when one looks beyond the rhetoric of participation, one finds little space left for schools and communities to decide how to spend their resources. One professor of education explained:

[I[f decisions are made at the center it is very difficult for them [central government authorities] to make these kinds of different needs analyses. This also reflects on the district level. The district never questions what the central government says. So they just comply. We raised the question as to whether there is even space for the district to question the center. At the school level and at the district level, we asked if they ever write to the center to raise issues or say that things aren't okay. There are no ideas or concerns flowing upwards from the district to the center. We do what we are told to do. This is an aspect of a highly centralized system. (Interview, May 23, 2005)

Views such as these suggest that school committees and district officials are now "at the table," but the options offered to them are limited to those laid out in central government policy. There continues to be little flow of decision-making power from the local to the national, and certainly international, levels even though local policy actors have had more training and experience in making such decisions through recent policy reforms.

Concluding Thoughts

The quotations that begin and end this chapter highlight the importance of the space that policies such as PEDP provide for broader engagement in education governance. The opening statement by HakiElimu's former director recognizes the space provided by PEDP for communities to participate in education governance; the professor quoted in the final excerpt, how-ever, questions the confines of this space wherein citizen engagement is extremely circumscribed. Their views indicate the potential for *ushiriki* but also the tendency toward *ushirikishwaji*. By attending to experiences of participation at various levels, this chapter reveals that PEDP, indeed, enrolled a broader range of participants than past policy reform initia-tives. However, the keyword participation promises more than it delivers: By and large, the "local" participation supposedly afforded through the policy is restricted by the centralized educational system. And yet, once enrolled in the network, parents, community members, and other partici-pants can utilize the policy, as an object-actor (see chapter 1), to expand their influence over educational decision making, especially when tutored by an organization such as HakiElimu.

The participation engendered through inter/national educational pol-icies in Tanzania is not unlike the participation stimulated by a leftist municipal government in Brazil (see chapter 5). In both cases, new policies opened spaces for the inclusion of grassroots policy actors, and yet well-established centralized organizational structures (in the case of Tanzania) and clientelism (in the case of Brazil) curbed the promise of participation. In contrast, the networks described in chapter 6 by Hantzopoulos, ini-tiated and sustained by teachers and administrators to implement their vision of participatory schools, seem better situated to avoid the challenges of *ushirikishwaji*, or participation orchestrated from above. By attending to discourses and experiences of participation at various levels in Tanzania, the complexity of this keyword becomes apparent and calls for us to rec-ognize that sweeping accolades and resolute criticism are inadequate. The vertical approach to policy studies used in this research allowed me to access multiple perspectives on participation to show that a policy such as PEDP has its faults but has created an opening for greater local input in decision making. Nevertheless, there is still a long way to go to get from *ushirikishwaji* to *ushiriki*. We need to continue to examine how policy actors and development professionals are ensuring that citizens effectively work within available decision-making spaces and how citizens themselves are attempting to widen them.

Notes

1. Haki elimu means "educational rights" in Kiswahili, the national language of Tanzania.
2. HakiElimu mobilized CIVs in Ukerewe and Serengeti districts to ensure that information on educational and community governance issues became available

in their communities. They maintained community and school notice boards, distributed publications, and encouraged school and village leaders to share information (on budgets and decisions) with citizens.

3. Documents reviewed included Consultant's Report: Evaluation of Human Rights Training in Serengeti, Bunda, and Ukerewe; Rural Development Policy; Ukerewe District Council—Progress Report on Child Survival, Protection and Development (CSPD); Latest Progress Report of Poverty Alleviation Initiative Developers (PAID); Terms of Reference for the Joint Review of PEDP 2003; Work Plan and Budget, July 2004–June 2005, District Development Program; Training Materials on Procurement in Local Government Authorities: Notes and Case Studies; and the Primary Education Development Plan (2002–2006)—Annual Performance Report for the Period of July 2002–June 2003.

4. By December 31, 2004, the percentage of required subdistrict meetings had significantly increased since the inception of LGRP in 2000. The occurrence of village council meetings increased from 30 percent to 60 percent; village assembly meetings increased from 25 percent to 40 percent; and Ward Development Committees increased from 25 percent to 75 percent (URT 2005a).

Chapter 5

Living Participation
Considering the Promise and Politics of Participatory Educational Reforms in Brazil

Moira Wilkinson

Um outro mundo é possível/Another world is possible.

—World Social Forum motto, Porto Alegre, Brazil 2003[1]

Porto Alegre, Brazil established itself on the world map of leftist-leaning politics in January 2001, when the city hosted the first annual World Social Forum. That inter/national social movement grew from a desire among radical activists and academics to critique the World Economic Forum that was taking place at the same time in Davos, Switzerland among the Group of Eight countries. The intention was to create a space where those countries excluded from the economic summit could discuss alternative routes to global political justice and economic stability. Inclusion was a goal of their efforts. Popular participation was the means to achieve it.

By 2001, participation was, in fact, a keyword firmly embedded not only in the working vocabulary of actors in that movement but also in the discourse of those politicians and development policies the activists struggled against as well. It is not surprising that these opposing groups of actors should agree on one premise: Increasing participation is, intuitively, an appealing prospect. However, while both sides have celebrated the increasing numbers of people who participate in democratic social experiments and policy reform, insufficient attention has been paid to *how* people participate and the quality of their experiences. In this chapter, I argue that the actual conductivity of participation is heavily contingent upon how participatory discourse is appropriated by the actors involved and by the behaviors, practices, and interactions that take place among them in participatory venues. A vertical case study approach allowed me to examine the very locally situated democratic experiment in Porto Alegre within national discussions of decentralization of the Brazilian state and alongside similar discussions among Northern international development agencies seeking to expand democracy. Within five years of its establishment, news of Porto Alegre's participatory democratic experiment reverberated internationally as a new model for democracy building and transformative social justice. After 16 years of this practice, however, my study reveals that cultural politics continue to impede the rich promise of participation.

Why There, Then?

It was not a coincidence that Porto Alegre was seat of the World Social Forum, nor was it arbitrary that schools figured centrally in the city's approach to increasing citizens' political participation. The city government, led by the Workers' Party (*Partido dos Trabalhadores*, or PT), had for a decade infused multiple opportunities for popular participation into city politics. Specifically, the city initiated an extensive and complex Participatory Budgeting process that intended to provide a channel through which poor and working-class citizens could make important decisions about expending scarce resources. The budgeting process became an inter/national reference for participatory politics.

Nested firmly within the PT's vision, public primary schools in Porto Alegre served as repositories of and incubators for the Party's democratic goals through the Citizen School project, which restructured the schools' administrations into site-based management teams called *Conselhos Escolares* (School Councils). The project has been described by its supporters as a "national and international reference in the construction of a school system where the right to knowledge [was] not a privilege of the few" (Gentili 2000, 17). Through mechanisms such as School Councils and Participatory Budgeting, the Workers' Party administrations that took shape at the end of the country's 30-year dictatorship sought to define how decentralization, coupled with substantive participation among a wider swath of Brazilian citizens, could create a democracy in the people's image.

To fully appreciate the revolution that the Workers' Party in Porto Alegre represented at the end of the 1980s, and to imagine why their vision captivated citizens and academics from many countries, it is important to understand the overarching national political culture *from* which the PT's municipal experiment emerged, as well as the immediate political, economic, and social context *into* which the Party inserted its model. Carvalho (2002), a Brazilian historian, synthesized the origins of this national political culture:

> Upon proclaiming its independence from Portugal in 1822, Brazil inherited a civic tradition that was hardly encouraging. In three centuries of colonization (1500–1822), the Portuguese had constructed an enormous country endowed with territorial, linguistic, cultural, and religious unity. But they had also left an illiterate population, a slavery-based society, a monocultural and feudal economy, an Absolutist State. At the beginning of the era of independence, there were no Brazilian citizens, nor a Brazilian nation. (18)

The legacies of the skewed social, economic, and political structures existing at the time of Independence were deep and long-lasting. For decades, representation was limited to a very small percentage of the population comprised primarily of the land-owning class and educated elite; national decision-making power was concentrated in the hands of a few. The combination of widespread illiteracy and wealth disparity created ideal conditions for political manipulation. In the eras of representative government, politicians routinely exchanged goods or political favor for votes. This practice of patronage, or clientelism, effectively reduced democracy to a superficial economic transaction and fostered little institutional or individual accountability. In fact, between 1822 and 1989, Brazil's relationship with democracy was tenuous at best. In those 167 years, the country had more experience with nondemocratic forms of governance— monarchy and dictatorships specifically—than with democratic ones. And even the periods that were technically representative were characterized by charismatic and authoritarian presidents such as Getúlio Vargas (1930–1945, 1950–1954). These nonrepresentative regimes, along with brief intervals of democracy, all reinforced the simple division of citizens into one small group of powerful elites and another comprised of everyone else (Carvalho 2002).

The absence of historical examples of representative democracy on the national level opened a space in the political imagination of the PT to surpass traditional conceptions of democracy and to rupture with the extant political culture of patronage. The Party's idea to transform democracy from within did not just fill this political void; it also responded to the serious economic disparities that were part of Brazil's inception and that had been protected, often violently, by its political institutions. The PT saw participation as the catalytic variable to economic, social, and political revolution because greater participation ensured equity, accountability, transparency, legitimacy, and ultimately justice.

The PT confronted a host of problems from the previous era. These included, most notably, the concentration of political, economic, and social resources among the elite and the distortion of representative government by favoritism, institutional opacity, and unequal resource distribution. The Party sought to create a deeper democracy by inverting the traditional priorities of government through expanding participation, increasing transparency, and serving the interests of the poor and working classes first. Schools were a critical site for this transformation. The Workers' Party's participatory response to the long-term political "crisis" in Brazil was to formulate policy that acknowledged education, in itself, as a basic right and, further, that positioned citizens armed with key knowledge and skills as significant participants.

Workers' Party's Inversion of Priorities for Inclusion and Democratization

> The popular support and participation [of common citizens] ensures the priority for social inclusion, for the improvement of the quality of life, principally for the population that lives in the periphery of the city... It is necessary to emphasize that the process of democratization of the State, the inversion of priorities... make[s] concrete the advances in the construction and conquest of citizenship. (Wampler 2003, 60–61)

The "inversion of priorities" is the ideological and operational emblem of the Workers' Party's vision of democracy. In contrast to more conservative practices of democracy, the phrase reflects the positioning of the poor and working classes, and not the landed elite, as both the target beneficiaries of decision making and its agents. Tarso Genro (1997), the second PT mayor of Porto Alegre, characterized the critical participatory features of their project as the "socialization" of politics. Specifically, he noted that the PT's model systemically "socialized" politics in three ways: (1) by creating public venues in which a diversity of citizens could make public their private interests; (2) by enabling citizens to relinquish less of their decision-making power in fewer spheres of their lives; and (3) by facilitating opportunities to "learn" citizenship through doing it. Embedded in the Party's approach was the assumption that participation alone was potent enough to activate a chain of associations linking participation to inclusion, citizenship, and democracy. This assumption was consistent with the dense inter/national discourse about participation, the procedural underpinnings of democratization, and the power of underlying political cultures. These three concepts form the framework for the analysis below.

The most common rationale for introducing participation into political mechanisms or development projects is to promote inclusion. As Linz and Stepan (1996) argue, "[T]he chances of consolidating democracy are increased by state policies that grant inclusive and equal citizenship [and that provide] a common 'roof' of state-mandated and state-enforced individual rights" (26). However, much depends on what form those policies take and how relevant actors appropriate them. Critical proponents point out that the forms and boundaries of participation are typically circumscribed by those in power and may be limited to what that group is willing to cede (Rahnema 1992; Kapoor 2002). White (1996) (see also Cleaver 1999) further points out that participation is, at its core, political in nature. Indeed, who participates, how, and on whose terms are questions inherently embedded within the existing power relations, or the cultural politics of the larger society (Alvarez, Dagnino, and Escobar 1998). These cultural politics strongly influence how people with and without socially ascribed privilege interact in mixed groups and navigate public spaces.

Ubiquitous cultural politics shape people's sense of entitlement to a place, and those places—schools in this instance—are not empty of meaning. The schools in Porto Alegre had preexisting political cultures into which citizens who got involved with the Party's reforms inserted (and asserted) themselves. The reverberations between the political culture and cultural politics affect what happens in institutions like schools (Alvarez, Dagnino, and Escobar 1998), and the potency of the experience therein can attract or repulse further interaction and shape conclusions about the value of participation and democracy.

Schools were deliberately selected by the PT as a site to interrupt Brazilian political culture because of the real and symbolic significance of these institutions in the public imagination. Raul Pont (2000, 29), Porto Alegre's third Workers' Party mayor, made the connection among participation, schools, and social transformation explicitly:

> This participatory democracy has as its instruments the Participatory Budget, Municipal Councils, School Councils, the most varied forums of direct, community participation and the use of direct consultation with the population.... [Participation is] the central element: beyond democratization, transparency, it is the link that connects the strategic struggle to overcome the limited criteria of representation that dominate our political system today.

It was hoped that, through participating repeatedly in these new spaces that were officially tasked with decision-making authority regarding allocation of public resources, and by prioritizing which values, skills, and knowledge to offer their children, the participants in this democratic process would form a sort of muscle memory of active citizenship and gain confidence in creating a society in the image of the majority of previously marginalized Brazilians. The Party's intuitive and appealing vision in Porto Alegre hinged on the promise of participation. Its success, however, depended on the actual conductivity of the emerging democratic spaces to produce equity, transparency, accountability, and justice in ways that traditional models of democracy had not.

Site, Methods, and Research Participants

The PT's goal in Porto Alegre to "invert priorities" was not only one of economic redistribution but also one of political representation. In the public mind, schools represent the repository of values to be passed on to the next productive generation of workers and citizens. This orientation plays out in policy and planning as well. Because of the ideological importance of schools, and education's explicit centrality in the Workers' Party comprehensive participatory experiment, during one phase of my overall study I chose to examine the participatory culture within the School Councils.

While selecting neighborhoods and then schools to study, I privileged sites whose conditions were most favorable for the consolidation of alternative democratic practices to get the fullest picture of what participation could make possible for democracy. I sought a neighborhood that had, for instance, a larger than average proportion of poor and working-class residents; a history of civic participation predating the return to democracy in 1985, the start of the democratic era; and a strong and consistent presence in the citywide participatory forums in the previous 15 years. Based on these criteria, I selected the region of Restinga, located on the outskirts of the city.

Within Restinga, I observed and interviewed participants in the School Councils of three of the seven public primary schools over a period of ten months. The councils were comprised of the various constituencies in the school community: administrators, teachers, parents, and students. In theory, the number of School Council members from each group was roughly tied to the size of the school and could expand up to 18 members (for more on the organization, structure, and origins of School Councils see Gandin 2002). In practice, the meetings were usually only attended by the administrator and the parents. Attendance at the meetings averaged six participants and no meeting had more than ten participants. Within the councils of the schools participating in my study, I focused exclusively on parents' constituency. For reasons related to gendered employment patterns and division of domestic labor, responsibility for children's education in Brazil is seen as a mother's responsibility; therefore, all of the School Council parental representatives were women. Each school had roughly the same number of students and was, therefore, allotted the same number of parental representatives (three) for their School Councils. Over the course of the school year, I attended 18 monthly School Council meetings across the 3 schools and conducted semistructured interviews with each of the 9 School Council parent representatives. I also conducted observations in the schools and informal interviews with parents not on the council.

Though I mostly consider data drawn from these sources, this chapter is informed by the larger study. During this period, to consider another site of democratic education, I was also studying the participatory budgeting process throughout the city. In addition, throughout the research process I attended neighborhood meetings discussing relevant social issues outside the official school sites. More generally, I engaged local social scientists and educationalists studying participatory processes, political officials representing the national and state parties, and current officials in the mayor's office. I also spent considerable time in the Workers' Party and municipal public archives examining documents about the policy reform process to construct the fullest picture of the Party's aims and vision and to create a backdrop against which to measure the participatory realities of the participants in my study.

Construction of Participatory Culture

The School Council is the maximum organ of the school. It brings together representatives of the parents, students, staff, teachers, and administration of the school, constructing a permanent channel of democratic and participatory practices in the consultative, deliberative, and fiscal aspects. It is the management organ that creates mechanisms of effective and democratic participation in the school community, from the definition of programming and application of financial resources, to the political-administrative-pedagogical project, to the elaboration and alteration of the school rules, to the definition of the school calendar, and enforcing the law. (Secretario Municipal de Educaçao [SMED] 2007)[2]

This statement from the Web site of the PT-headed Municipal Secretariat of Education delineates the fullest interpretation of what the School

Councils were to achieve within the Party's vision. It is clear that the Party strongly advanced the idea that participation in key political places, such as Schools Councils, could promote inclusion, citizenship, and democratization. To accelerate the social transformation they envisioned, the Party linked participation on the councils to meaningful democratic processes and weighty content; what Gandin and Apple (2002) called "thick" democracy. However, their goals were subverted by the tenacity of existing political cultures of school administrations and the cultural politics between the administrators and the people they served. Contrary to the earlier stated goals, the conduct between the administrators and parents serving on the councils during the School Council meetings was hierarchical rather than egalitarian, and the content was limited to verifying budget and expense reports prepared by the administrators. Further, the nature of the interactions between the School Council members and the administrators between meetings showed that the same type of preferential treatment that permeated traditional political institutions throughout Brazil had been reproduced inside the transformative venues. In this section, I illustrate these cultural dynamics in both the monthly meetings and the everyday encounters between the administrators and the School Council volunteers in the weeks between meetings.

Monthly Meetings

The uniformity of the meetings across the three school settings was striking. Although the mission of the School Councils was to influence fundamentally both the way the school was managed and the type of knowledge produced there, only 1 of the 18 School Council meetings I attended addressed a topic relevant to these purposes. Meetings were not sites of thoughtful, animated exchange about critical content and pedagogy. Instead, they were limited to rote explanation by the administration of the necessary purchases for the school. The meeting described below was typical:

> 1:55 pm: The Principal explained that she needed the Council to look over and initial the expenses from the month before. She listed briefly what they were related to. There were over 15 documents that looked like tables and were itemized according to the category of expense to be signed by each of the people on the Council. There were three packets with one set of documentation each. One had original receipts attached to each of the invoices and check stubs showing the number of the check that the Principal had used to pay the bill. All of the checks were filled out and signed by her. She told the Council Members to look them over carefully, that each of

the three copies had to be filled out by hand and that if there were any discrepancies or if anyone signed in the wrong space, they would have to re-do them and sign them again. As these documents passed from one person to the next for signing, occasionally a Councilor or the Principal joked or reminded people not to mix them up because of the problems it would cause, and they laughed. The Principal also interjected with an occasional complaint about SMED and all the paperwork they had to do. The women laughed at this too, or made sympathetic remarks about how little money there was to spend in the first place. Shortly after they were done, about 2:25 pm, the meeting ended. (Fieldnotes, August 3, 2005)

Solange, one of the parent councilors from this school, described the purpose of these meetings from her perspective:

We approve or disapprove of what they are going to spend the money on. They receive funds every 60 or 45 days, something like that, and then they show us the planning for six months, I think. I don't know if it was even six months and they said, "In these six months, we're going to buy this, this, and this with the money, or we have this and this to buy. What do you think we should buy first?" And so from there we approve it and they can't buy anything without our signature. (Interview, August 3, 2005)

The meeting described earlier, as well as Solange's explanation of the purpose of such meetings, exemplifies the limited participatory culture in the School Councils. In particular, the duration and content of the meeting and the behavior of the participants raise questions about how democratic these meetings were in practice. School Council meetings rarely lasted longer than 45 minutes. The principals and not the participants controlled the content: they set the agenda and the tenor of the meetings. The behavior of participants was hardly what one might imagine would constitute a lively display of democracy in action. Instead, the women primarily approved the decisions already made by the administrators. Notably, the principals were also the fulcrum of the exchanges among participants as they rarely addressed one another directly. Due to the infrequency of the meetings, coupled with the principals' regulating behavior when they did meet, the female councilors formed little or no community cohesion among themselves. Instead, the relationships that did emerge were individual ones between a single councilor and the principal.

In light of the consistency at the meetings and the divergence from the prescribed participatory goals for the councils, I asked each female council member if she had ever participated in a meeting that stood out in her mind as different from the ones I attended. Five of the nine women could produce only one such example, and one councilor, Solange, mentioned two. One of Solange's examples had happened the year before. In that meeting, the school psychologist had suggested to the councilors that they form a group to visit the homes of students who consistently came to school unbathed and unkempt. Solange was opposed to the idea because she thought it was intrusive into the lives of these families, but she cited the meeting as one she liked because the substance was rich and required deliberation among council members. Like Solange's example, the recurring theme in the examples of unusual meetings by the five other councilors was that the topics of discussion were substantive and related to educational issues. The overall scarcity of examples from the women's one to two years of experience—spanning five to twenty meetings each—suggested that meetings with weightier content were aberrations rather than the norm.

The enthusiasm with which these council members related such anecdotes was uncharacteristic of their descriptions of regular meetings. The women had found value in those unique interactions, increasing the appeal of service on the School Council. One woman explained it simply, "Oh, I like it that... talking, and I get to learn, because everything for me is a lesson... I keep learning and sometimes, I just listen, to learn. We are learning, you know" (Interview, September 28, 2005). In fact, at one point or another six of the nine councilors echoed these sentiments. Even though very few of the meetings were substantive, the councilors viewed such meetings as a gateway to learn what was going on in the community and to be informed. The women's sentiments about these meetings simultaneously corroborated the Workers' Party assertion that participation would lead to more intense civic engagement and underlined the missed opportunity of the School Council meetings by generally limiting the content to budget approvals.

Though most of the female councilors responded excitedly to making decisions of importance, in the 18 meetings that I attended, none of them ever introduced a topic.[3] This demonstrated further the lopsided nature of the administrator-councilor relationship. When I probed about this topic during interviews, asking what they *might* like to discuss at meetings, only Aparecida could identify something concrete. Her answer was revealing. She was both a mother and a student herself in night classes at the school. The citywide *Escola para Jovens e Adultos,* or *EJA* (School for Young People and Adults), was the public school secretariat's alternative program for overage teens and returning adults to finish primary school, and it encompassed Grades 1–8. At her school, however, EJA was only offered up to the third grade. She had completed that grade twice already (the second time, she said, because it allowed her to get out of the house), and said she accepted the invitation to run for one of the parent seats on the council because she was hoping to get them to expand to the fourth through sixth grades so that she could finish her basic education. As Aparecida explained, she nonetheless did not raise this desire in the meetings:

> Aparecida: I think it doesn't have anything to do with the meetings.... [I]t was what made me enter, what gave me most strength to enter, but I don't think I understood the thing, because I thought, having the council, you know, being a councilor we could get the next grades to come here, that we were going to fight for that.
>
> Moira: For what?
>
> Aparecida: EJA, night school, but from the meetings that we've already had, I guess that it doesn't have anything to do with that, it wouldn't do any good to even bring it up.
>
> Moira: Why not?
>
> Aparecida: Because it hasn't been raised until now, you know? And I wait until they touch on a subject, and I only see, so far, that it's only school costs, the expenses, you know, what there is to fix, so I think that it's something that.... The councilors, they aren't going to be able to do anything about it. It was that that got me excited about entering. (Interview, September 27, 2005)

It is notable that Aparecida thought she was wrong about what was supposed to happen at School Council meetings. She was explicit about the fact that whatever was done in the meetings was how this putatively democratic space was supposed to be used; she did not question the authority of the principal in setting the agenda and, instead, discarded her own notion of what the meetings should be like. The pattern of principals setting the agenda did little to ameliorate the already wide gap these women perceived between themselves as community representatives and the administrators as government officials.

The deference the women showed toward the administrators was evident, too, in the disjuncture between what happened in the meetings—the way the councilors acted in meetings—and the way they portrayed the meetings to me later. Even the single atypical meeting I attended underscored the durability of the authoritarian political culture of the School Councils. The weekend before the unusual meeting, there had been an extracurricular program called *Escola Aberta* (Open School) at the school, and two men were employed by the administration to oversee the security of the children and the property while the school was open. One of the men in charge was employed as a school security guard during the week, and the other was a father who participated in the EJA program at that school and represented the student constituency on the council. During *Escola Aberta*, some students were found vandalizing the building, and the two men had to discipline them. The program was new, and staffing issues had not been fully formalized. A difference of opinion between the two men in charge about how to discipline the youth who were misbehaving escalated into a heated argument between these two adults, each seeing the intervention of the other as undermining his authority. In the council meeting days later, the council discussed the men's eruption. The altercation between the security guard and the father had not been resolved at the time it happened, and the question about the person who had primary authority remained unclear. The issue was brought in front of the council, apparently for mediating purposes. In interviews with councilors from that school later in the week, I took the opportunity to ask them about their interpretations of what happened at the meeting and the issue's resolution. Aurora, a female councilor, offered the following account two days after the council meeting:

> Aurora: The assistant principal and the principal invited us, me, Matilde and that guy, to…how is it? Participate in *Escola Aberta*. We felt bad for him, because he was unemployed, father of a family, you know, he was studying at night, EJA… [She hesitated.] So, we let him, you know, and the first thing he did was to fight with the teacher…Sandro [the security guard] was left in charge, in general, of the school [during that program]. He [Sandro] had the key, and the other one [the adult male student/councilor] was in charge of…of doing the maintenance, keep[ing] an eye on what was happening in the courtyard, open[ing] the gate to let people in, you know? But he [the male councilor] thought that no, that he was the one running everything, he wanted to go over Sandro's head and Sandro didn't accept that. They fought badly, you know, so ugly that in the meeting he even fought with us and I said, no, a person like you can't be in charge in a place like that, even more public, where two of them were fighting, he takes everything, like, *a ponto de faca* [at knifepoint, or by force] you know? They had a…how would you say it…you know right then…how do you say it? A vote, and the majority asked for him to leave [the premises], so he left. (Interview, November 18, 2005)

According to my fieldnotes (November 16, 2005), the meeting had publicly shamed the councilor in question and was less democratic than Aurora's summary suggested, even by normal standards of School Council meetings. Because I had been at the meeting and Aurora was aware of this fact, it is particularly striking that she was initially ambiguous about what happened. Her reluctance, at first, to discuss the meeting indicated that she was not comfortable sharing what had occurred. More specifically, a topic such as this one regarding, at its base, employment, had never before been brought to the council, as far as I could tell. Although employment matters were within the purview of the council's mandate, many councilors felt that asking the man to resign his seat because of perceived misbehavior was misaligned with the purposes of the council itself. According to the other participant I interviewed after the meeting, the principal had already discussed the incident informally with many of the members before the meeting and had already given an ultimatum to the councilor to acknowledge his error or leave the council. Bringing the matter to the council, then, was not for purposes of mediation, because in the principal's mind the decision had already been made; rather, the strategy appeared to be one of applying public pressure on the father/student to do as told. Though some of the teachers spoke up to defend him, there was silence among the parent councilors, and they voted to have him dismissed. In the end, the consequence was dearer still because he not only left the council but his school program as well. This example and others show that the expectations regarding behavior between individual councilors and school administrators were vestiges of entrenched antidemocratic behavior; it will require more than participating in School Council meetings to overcome them.

Mundane Encounters

Council members' compliance with figures of authority was often compensated in the informal interactions among them and administrators, further subverting the democratic purposes the School Councils were intended to serve. Unlike other parents of students at the schools, each of the parent councilors gained privileged access to school grounds. They reported that they were able to enter the school compound (always locked to the public during school hours) during parties that were usually closed to the other students' families, though their presence at these events was not formally related to being on the councils. When I asked Solange what benefit she got from her participation on the School Council, she gave the following example:

> My kids love it [that I'm on the council], you know, if I have to go to school with them, have to do something. Like when there's a little party, you as a mother from the council can enter. Generally for the other mothers, it's closed, you know, because if not, there would be chaos. In the *Festa de São João* [St. John's Day party], I go over early, I cut the cake, we make popcorn and fill bags with it. The whole day until the party, you can help. I think that's what you gain from it, there isn't anything else, and I think that's the best thing...Before when they were a little younger, I even went on the field trips, the principal always invited me, sometimes I even went on trips that weren't even with my kids' classes. (Interview, August 3, 2005)

This literal gate-keeping by the principals directly contradicted the PT's message that public schools belong to everyone. First, the locked entry to the schools—except for parents on the School Council—created a real barrier between the public and their schools. The PTs' vision to have parents shape their children's knowledge became impossible because they could not get in the building. Second, councilors received rewards for which other parents were not eligible, with some of the most important being access to school personnel and privileges for their children.

Some of the women reported gaining personal and financial benefits from these council-based contacts, but by and large, the mother-councilors used their heightened status to gain attention for their children. One day, for example, as I was standing with a principal, Amélia, a female councilor, approached and requested a certain teacher for her son (Fieldnotes, June 21, 2005). Likewise, councilor Aparecida was often at the school asking about her child, who was frequently truant or in trouble with his teacher. After getting in the gate because of their status as councilors, they enjoyed access to the administration to ask for favors. The most dramatic instance of preferential treatment came in the form of side-stepping enrollment protocol altogether for councilors who had children at more than one public school and who wanted to bring the others to the school where they participated on the School Council. Vanessa's observation as a councilor powerfully but subtly conveys this sense of privilege: "If my son has a problem, I can go straight to the Administration, you know? I can go right there whenever I want and talk with them, you know? I have that freedom with them and I feel good. The teachers, the majority, they know me and they come and say hi and give me hugs" (Interview, November 18, 2005). Similar to Bartlett's (2007; 2009) study of Brazilian literacy programs, in which Freirean teachers' "friendship strategy" in class unintentionally subverted the more radical democratic agenda, the parents in this study used their position on the council not to further democratic participation but to form relationships that secured benefits for their own children.

Ideally, School Councils were supposed to democratize educational management and knowledge production in schools. According to my observations and the female councilors' accounts, the reality diverged widely from this goal. Instead, the scope of the contribution the women made in their capacities as parental representatives was limited to endorsing decisions by administrators and signing forms. The procedural and behavioral uniformity alongside the absence of conflict during the School Council meetings would not have been found in an environment of debate and egalitarian exchange. The formulaic nature of the meetings, the tacit power held by the principals, and the women's own desires to make requests but their reluctance to do so spoke rather to a very limited culture of participation and "thin democracy" (Gandin and Apple 2002).

Perhaps more importantly, the contribution parent councilors made was precisely the contribution they were allowed to make by administrators. This is similar to the situation described by Taylor in the previous chapter in which Tanzanian community members were not choosing policy options from a "buffet" but rather off of a limited "*prix fix*" menu. Moreover, these Brazilian mothers showed few signs of representing the community's interests at the meetings, much less deliberating over vital educational decisions. The preferential treatment the parent councilors received from school administrators demonstrated that old political practices carried on in new participatory venues. What the women shared in their interviews led me to conclude that participation on the councils had not enabled them to fully assume the role of School Councilor as envisioned by the Workers' Party, or even as fully as the councilors themselves would have liked. Furthermore, their frequent silence during meetings suggested that the cultural politics among the council members played into uneven power dynamics in Brazilian society more broadly. For the most part, council members occupied their expected class-defined roles, leaving the existing political culture uninterrupted.

Reconciling the Promise and Politics of Participatory Democracy

> The lessons of democracy have been incorporated into everyday practice, permeating the relationships of power that are developed inside schools and in its interface with the community. This relates to the school producing, within the educational processes and in the set of relationships therein, methods and practices that have implications through their contribution for the formation of a democratic culture, capable of diffusing itself throughout the social body. (Azevedo 2000, 69)

Azevedo was one of the city's Secretary's of Education appointed by the Workers' Party. The earlier quote appears in a book edited by him on the education approach of the Party and its effects on the city's schools under the PT's consecutive city administrations. The image Azevedo created is an inspiring one that ought to guide educational innovators. Unfortunately, contrary to Azevedo's contention, the dynamics on the School Councils I studied did little to genuinely democratize the Porto Alegre public schools. That practices of participation existed in the city was evident; however, the way in which the inter/national discourse of participation was appropriated by the Party's municipal leadership and the school administrators—and not taken up by the councilors—does not support the view that School Councils were more democratic than the traditional school governance the Party sought to transform. The closed nature of the schools and historical authoritarian relationships that favor certain groups, such as middle and upper classes and the educated—dynamics the PT eschewed—were reproduced on the councils as well. School councilors got privileged treatment and received private benefit for the public work they did for their communities. As O'Donnell (1996) cautioned, the existence and appearance of democratic processes often mask persistent and familiar antidemocratic behaviors. In the case of the School Councils, the resulting political culture subverted their full democratic spirit and capacity and diverged markedly from the portrait advanced by the PT.

The Workers' Party's most vital assumption was that the conduct in the Party's participatory governance bodies was egalitarian and deliberative enough to usher in a new, deeper, and more inclusive kind of democracy. That is, they believed that the conductivity of participation was sufficient to invert the traditional social order. However, this study shows that female councilors usually waited for direction from the administrations and never advanced an agenda topic of their own or voiced disagreement with the administrators. These observations and interviews contradict the conclusion that such opportunities to participate cultivated fully participatory cultures in a range of social spaces or that participation by itself was a sufficiently powerful vehicle to achieve the social and political goals the Party intended. In reality, the potential for this experiment in participation to achieve social inclusion and democratic consolidation in Brazil is complex in ways that are not customarily acknowledged. Despite the progressive vision, painstaking planning, and intensity of resources the municipal government dedicated to this work over the years, the School Councils carved out a much narrower wedge than the PT imagined, leaving the locus of control firmly with the school administrators and causing the female councilors to question their own understandings of the procedures when they diverged from those of the authorities.

Similar to several previous chapters, this one has shown that participation is a potent keyword that defies simple categorization. The councilors in this study resemble, in many respects, the citizens of Ukerewe (see chapter 4) and the members of the *Université Cheikh Anta Diop* community (see chapter 2) whose participation in an inter/national reform process facilitated educational change, but only within a circumscribed political space. As limited as it may have been, the councilors' participation has the potential to advance a larger democratic movement as part of the Workers'

Party agenda. Forming a confederation of school councils, similar to the small schools movement described in the next chapter, might be one way to promote a more radical shift in social relations and realize more fully the promise of participatory governance reforms in Brazil.

Notes

1. Quotation from the World Social Forum Web site: http://www.forumsocialmundial. org.br/index.php?cd_language=2&id_menu.
2. For details see the Porto Alegre prefecture Web site: http://www2.portoalegre. rs.gov.br/smed/default.php?p_secao=98.
3. Although the councilors did not put forward topics, this was not for lack of ideas. All of them voiced universal distaste for the "Cycle" format of the primary schools. The Cycle was introduced by the Workers' Party Education Secretary as a way to minimize failure and retention caused by children's different learning paces. In practice, the Cycles were seen to lower expectations for students and did not provide sufficiently rigorous mechanisms to identify and correct learning gaps before they became critical.

Chapter 6

Transformative Teaching in Restrictive Times

Engaging Teacher Participation in Small School Reform during an Era of Standardization

Maria Hantzopoulos

Our task [in the small schools movement] was to reinvigorate intellectually and professionally the educators who had survived typically 20 years in these anonymous and disempowering institutions, to re-engage the students "left behind" who attended (and did not attend) these schools, and to organize the parents who were attached to (and alienated from) the institutions of deep despair and not infrequent hostility.

—Fine 1994

The epigraph by noted education scholar Michelle Fine suggests that the small schools movement, an increasingly popular educational reform initiative in the United States, was originally conceived to expand the democratic involvement of teachers, students, and parents in their own educational experiences and in the creation of inclusive and innovative schools. However, the proliferation of small schools in recent years has resulted in many of the principles of the original movement, such as democratic participation, critical pedagogies, and project-based assessment, becoming obscured by an increasingly standardized educational agenda. As the privatization and commodification of education through government mandates, standards-based reform, and high-stakes testing (described in chapter 1 by Koyama) have become progressively more common practice in the United States, these changes have minimized the essential features of teacher, student, and community participation in school administration that were pivotal to the original small schools movement (Fine 2005; Hantzopoulos 2008; Klonsky and Klonsky 2008). Consequently, the small schools movement has been radically reconceptualized and ultimately bifurcated into different directions: one that emphasizes the democratic, participatory, and self-governing nature of the original movement that resists standards-based reform; the other that reflects and coincides with the trends in privatization, less local autonomy, and increased federal control over student and teacher assessment.

This chapter, similar to those by Taylor and Wilkinson in this section, explores the keyword *participation* but situates it in the United States as part of a vertical case study of teacher participation and engagement in the small schools movement. It explores how teachers at a small, collaboratively run high school in New York City navigate the contested terrain of centralized educational reform by becoming involved in professional networks of like-minded educators. Similar to Koyama in chapter 1, I examine the ways that national initiatives such as No Child Left Behind (NCLB) impact the lives of teachers in specific schools. However, my lens is focused on how one group of teachers attempts to maintain the participatory nature of the original small schools movement by looking at the ways that they participate in broader interschool networks to oppose standards-based mandates rather than succumb to them. I specifically hone in on educators at Humanities Preparatory Academy (HPA), an older small public high school in New York City where teachers strive to enact democratic, participatory, and socially transformative approaches to schooling, despite challenges that sometimes prevent this from happening. Drawing from one year of fieldwork at HPA that includes interviews with and observations of teachers, I not only show why educators seek to be a part of this particular learning community but also how they organize in ways that might help them maintain what they perceive as the unique participatory structures of the school. They pursue their goals, in part, by allying with kindred institutions to both strengthen and reflect upon their practices and form regional and national networks in opposition to government policies favoring standards-based reforms. Although democratic engagement and participatory action is fraught with contradictions and inconsistencies, I argue that through affiliating with various national and local professional development and policy networks, such as the Coalition of Essential Schools (CES) Small Schools Project (SSP) and the New York

State Performance Based Consortium (the Consortium), educators at HPA and other schools in these networks are catalyzed to confront restrictive mandates and, possibly, reverse them. Subsequently, educators forge new ground through these networks to engage in educational reform debates from which they were previously excluded.

These locally situated small schools, therefore, help shape the larger regional and national small schools networks, which, in turn, may contribute to redefining the small schools landscape in the United States and beyond. Through the involvement of teachers as participatory actors in school reform, these schools demonstrate the possibilities of preserving the principles of the original small schools movement in the current climate of continued neoliberal educational policies. To explore this process, the chapter employs a ground-up approach to the vertical case study, examining how linkages among local actors contribute to and inform broader regional and inter/national education movements. While the other authors in this section elucidate the limitations of participation in their chapters, I suggest that through *organically* created professional networks, teachers participate in school policy and practice in ways that help them navigate the enactment of their sui generis school missions in a climate of testing and accountability. Thus, while pure participation is never possible, these networks are generated by the educators themselves, creating a potential space for teacher negotiation, problem solving, and decision making in a time of hierarchical and standardized reform.

Small School Expansion in an Era of Standardization

The rapid proliferation and recent political support of small schools, particularly in New York City, had its roots in a movement that initially emerged gradually.[1] Starting with the founding of the innovative Central Park East School in Manhattan in 1974, several other pioneering small schools began to surface in urban centers during the 1980s and 1990s throughout the United States (Meier 1995; Fine and Somerville 1998; Clinchy 2000; Fine 2000, 2005). Although smallness was central to the creation of these schools, it was certainly not the only feature that defined them. These schools were also committed to personalized, participatory, democratic, and relevant education as a means for ensuring high levels of student achievement, particularly for historically marginalized young people (Apple and Beane 1995; Harber 1996; Nieto 2000; Powell 2002; Fine 2005).

Although each of the schools had distinct missions and curricula, many of them were inspired by the work and philosophy of Theodore Sizer (1984, 1992, 1996), founder of the CES. He believed that schools and classes ought to be smaller and that teachers, instead of bureaucrats, needed to run schools in partnership with students and parents, an idea that contrasted with the increasingly popular teacher accountability models that were surfacing in the 1980s and 1990s. These original small schools were committed to working on "authentic" tasks and projects dedicated to social justice with students;[2] accordingly, they were not simply about changing the size of schools but also about innovative curricular reform and establishing a more democratic and participatory approach to school administration (Fine 2005; Greene 2005). Multiple studies have pointed to the success of these small schools, particularly for working-class youth and youth of color, citing their much lower drop-out and higher college acceptance rates (Fine and Powell 2001; Hantzopoulos 2008; Rodriguez and Conchas 2008).

Many advocates of the original small schools movement, however, are now skeptical about the sudden exponential multiplication of small schools. They are particularly concerned about the ways that newer small schools have focused solely on size while ignoring the elements of democratic participation inherent in the original movement (Meier 2004; Fine 2005; Klonsky and Klonsky 2008). This aberration is largely due to the simultaneous advent of more centralized educational reorganization and high-stakes standards-based reforms. Culminating in greater government surveillance of schools and in legislation such as the federal NCLB Act and examinations such as the New York State Regents (as exit requirements), these reforms have removed a significant degree of local control for decision making by standardizing curricula, citing this as a means to ensure equal opportunity for all students and accountability for deteriorating schools. Further, these reforms have constrained more participatory approaches to the administration of schools as increasing amounts of time are dedicated to covering the mandated curriculum to the peril of other activities, including teacher collaboration and student-led meetings. As small schools continue to gain popularity and political support, original small school proponents warn that their mission as participatory, inclusive, and democratically run endeavors must remain central to their conception (Meier 2000a, 2000b, 2002; Fine 2005; Klonsky and Klonsky 2008).

Although the impetus to create a standardized core curriculum for the state and nation is ensconced in language about equity and opportunity, it also comes on the heels of and aligns itself with an increasingly accepted educational agenda that involves the private sector more actively in matters of public schooling (see chapter 1). As both standards-based reform and the small schools movement arose during an era of neoliberalism in the United States, there has been an increased role for the market in education, often backed and supported by government educational initiatives. Despite these claims of equity and fiscal efficiency, critics have argued that this educational agenda has only served to perpetuate and exacerbate inequality and deny opportunity for historically marginalized populations (Saulny 2004; Berliner 2005; Levister 2005; Gootman 2006b).

The shift toward market-based education is part of a larger inter/national phenomenon: the growing influence of neoliberal economic policies among industrialized nations that began in the 1970s and the eventual enactment of those ideas throughout much of the globe in the 1980s and 1990s (Burbules and Torres 2000; Raduntz 2005). Privileging particular ideas about how to operate the economy, these policies valorize economic prescriptions such as deregulation, free trade, and less government funding for social services because of their ideological faith in the market rather than in government (Burbules and Torres 2000; Morrow and Torres 2000; Raduntz 2005).

According to Burbules and Torres (2000), neoliberal policies not only provide recommendations on how to restructure the economy but also on how to transform education, politics, and culture. While the rhetoric of neoliberal reform decries expansive government involvement in education, government agencies have in fact expanded their role in this arena through their support of corporate initiatives in education. The government facilitates this privatization process as legislation, institutional forms, and authoritative discourses promote a reconstitution of public goods that favors the private sector's involvement in education matters (Bartlett, Frederick, Gulbrandsen, and Nurillo 2002). For instance, the private-testing industry is involved in not only the creation, preparation, and grading of assessments, but it also now provides remedial intervention in "failing" public schools as part of NCLB. As Koyama highlights, federal monies like Title I that are secured for schooling are *less* likely to directly benefit students since the enactment of this policy. Instead, this money is redirected to pay for after-school tutoring, often provided by test-preparation business entities. Yet, there is no substantial evidence, and no real accountability systems, to ensure that these private enterprises are improving test scores (Miller 2008).

It is not only student performance that is measured through standardized test scores, but also teacher professionalism and accountability, as educators are increasingly under scrutiny when they do not meet federal or state benchmarks. Scholars indicate that the standardization of teaching and curriculum demanded by standards-based reforms is closely linked to the de-skilling and de-valorizing of educators (Morrow and Torres 2000;

Singh, Kenway, and Apple 2005; Miller 2008). Due to this pressure for their students to achieve high marks on standardized exams, educators' knowledge is called into question as worker accountability, performance standardization, state testing, and core curriculum are used "in odd ways to reinforce each other" (Burbules and Torres 2000, 14). Tyack and Cuban (1995) assert that teachers in the United States, in response to intensified standards-based reform, rely less on their professional judgment and complain of more bureaucracy, paperwork, assignments, and stress.

The implication of this ideology for schooling is that as education becomes "marketized," it further advances notions such as individualism, competition, and consumerism (Raduntz 2005, 234). Quality education is treated as a commodity, rather than as a right, ostensibly giving parents a choice among schools as if they were selecting a product. Examples of this trend are the increase in charter schools, voucher programs for private schools, and tax credits for private education. Thus, policies that were once considered radically conservative have become "commonsense" educational policies (Apple 2000, 2001, 2005; Bartlett, Frederick, Gulbrandsen, and Nurillo 2002). These policies rest on the assumption that the market will ensure equity and justice; if schools are exposed as failing, then parents will choose better ones for their children. Yet, as Apple (2005) points out, there is growing empirical evidence internationally that the "development of [a] 'quasi-market' in education has often led to the exacerbation of existing social divisions surrounding class and race" (216). The work of Jane Kenway (1993), for instance, demonstrates how market-driven educational reforms in Australia deepen inequities in schooling and perpetuate educational stratification. Nevertheless, individuals still see their own interests as being served by the school initiatives that ideologically promote "choice," evident in continuing community debates over school vouchers, charter schools, and the role of business in school management (Smith 2004; Singh, Kenway, and Apple 2005).

The shape and direction of the original small schools movement has been profoundly redirected as a result of these neoliberal policies that emphasize hierarchical, marketized, and standards-based reform (see Fine 2005; Klonsky and Klonsky 2008). Now seen as the new panacea for educational failure, the small schools movement has partly become a vehicle for these policies despite the inherent paradox of school innovation and increased standardization. Consequently, many of the teachers/reformers who share the vision of the original movement believe that there is an exigent need to reclaim "small schools" and revive their democratic and participatory roots. Because many of these teachers eschew reforms that promote standardization (and by implication, de-value teacher expertise), several have joined networks that not only support their endeavors to enact their school missions but also provide a forum for activism against these types of reforms. Through their pedagogy, role in school governance, and participation in intraschool political work and social movements, these teachers are potentially cultivating what I call a *critical* small schools movement, distinguished from a movement of schools that are just small.

The remainder of this chapter, therefore, zeroes in on HPA, a critical small school in New York City, and its affiliate schools in the local New York Performance-Based Standards Consortium and the national CES Small Schools Network. I demonstrate how teachers, through alliances with other educators, participate in local, regional, and national organizations to oppose standards-and market-based reform and to present critical alternatives. Although studies often show how teachers appropriate aspects of policy to achieve their desired goals (Sipple, Kileen, and Monk 2004; chapter 8, this volume), I illustrate how they work together not only to preserve school missions but also to cultivate larger professional networks with the hope of interrupting standardized policies and practices.

One School at a Time: Growing a Critical Small Schools Movement through Teacher Participation

Originally designed as a school for students who were underserved in traditional school settings and were potentially at risk for "dropping out," the founders and subsequent teachers at HPA have sought to create a college-preparatory school that reengages students academically and emphasizes democratic principles of participation throughout its thematic and project-based curricula (Hantzopoulos 2008). To achieve this goal, the school relies on democratic staff collaboration, particularly in developing curriculum and in creating school policy. This type of participatory model aligns with much of the literature on school reform that indicates how participation in collaborative learning communities among professionals in schools will lead to both greater student achievement and overall school success (Wildman and Niles 1987; Lieb 1991; Senge 1991). Moreover, this form of professional development that emphasizes staff participation involves deliberately scheduling time for teachers to work together to design materials and inform and critique one another (Little 1982; Raywid 1993).

Although there are myriad reasons for educators choosing to come to HPA, many stated they desired a place that emphasized teacher participation in curricular and governance matters and valued teacher knowledge and expertise. By assuming that teachers (and students) bring assets to the classroom and school community, the participatory governance model at HPA departs from the deficits approach implicit in standards-based reforms. In fact, according to the school founder, Perry Weiner, the idea for the school was not only to counter an increasing disaffection with schooling that he noticed among his students but also to create a place in which faculty participate in creating imaginative and responsive structures for students. When asked why he started the school, Perry explained:

> I guess I intuitively seized the opportunity to create on a larger scale what I'd endeavored to do as a teacher in my own classroom: a humane school environment; an intellectual community; a place of mutually respectful discourse and some significant portion of democracy. Most of what I beheld around me in the larger school contradicted these. The hysterical obsession with school tone; the exclusively law-and-order approach to dealing with kids; the continual derogation of students, generally meaning (though not exclusively) Black and Latino students; the extraordinary lack of enlightenment in general...I had faith that a more hospitable leadership and structure would move students and teachers in the direction of greater humanity and greater achievement. (Interview, June 15, 2007)

By describing the culture of his old school, Perry explained how his hope was to create an alternative "hospitable leadership and structure" built upon teacher and student input. In this sense, student academic achievement was linked to staff (and often student) participation in the decision-making processes of the school.

Others who initially joined him in this endeavor expressed interest in establishing a collaborative learning community among adults. One founding teacher, Christina, recalled her previous teaching experience as one in which "when you could have a discussion [with other teachers], a few people held court and you were silenced. And talking about teaching was just unheard of" (Interview, June 5, 2007). She continued to say that this was partially "structural" within her former school rather than innate to the teaching profession. Thus, she was lured to HPA because of Perry's enthusiasm for radically restructuring and democratizing school leadership and practice. Similarly, Vincent, the founding principal, saw the potential of a professional learning community that could "play out more the linkages between democracy and education just on a philosophical sense and also in terms of my role as citizen...citizen/educator was how I wanted to think of myself" (Interview, March 19, 2007).

Thus, among the school's founders, there was a clear desire to establish an environment that emphasized personalization (over standardization), democratic, participatory structures for teachers and students, and faculty collaboration. At present, these desires are manifest in many structures at HPA, including weekly faculty meetings where staff members make decisions through consensus, often with student input, regarding curricula, assessment, and school policy. Moreover, there are twice yearly staff retreats to focus on self-reflection, creativity, and long-term teaching goals, all of which are co-designed and facilitated by the principal and a teacher. The following excerpt from an HPA science teacher exemplifies why this type of professional learning community might appeal to teachers:

> I spent my first three years in New York teaching in a pretty tough school... I was sort of coping with what I think is very typical in a lot of urban schools, which is the overwhelming concern with school safety, and turning around a failing school creates this culture of perpetual anxiety and creates a lot of really unfortunate thinking about kids. So the school I was in had very limited understandings of kids. They got their graduation rates up, they got their Regents pass rates up, but it was like the means by which that was happening were pretty atrocious, some of the stuff that was going on. So I hadn't had yet as a professional educator an opportunity to be with other professionals who thought what I thought and who were trying the kinds of things that I believed in. Everything I read about [HPA] suggested this was a place that was really taking risks and trying interesting things. That excited me—knowing that this place was governed by consensus... Seeing that Prep was so interested in engaging teachers and thinking about how the school ought to be run to me—at first anyway—seemed like a dream come true. Certainly since then I've realized that that mode of running a school has its own challenges, but all these things came together to me to say this is a place I need to experience. (Interview, May 8, 2007)

Most staff members at HPA expressed similar views about the high level of faculty input into decision making and curriculum construction, making these particularly attractive features of the school. Although many acknowledged that power dynamics among faculty sometimes interfered with individual staff participation in school governance, they also noted that the commitment to teamwork ultimately helped them enact aspects of the school's mission to resocialize at-risk students academically. According to Klonsky and Klonsky (2008), since teacher collaboration and respect for teacher knowledge was a pivotal practice of the early small schools movement, they suggest that schools cultivate these practices as a means to revive the (critical) small school movement.

Established communities of teachers, however, might not be enough to maintain participatory practices during a time of increased state and national pressures on schools to raise students' test scores. As noted earlier, curricular mandates potentially undermine the types of practices that take place in critical small schools like HPA. Such mandates not only lead to the standardization of curriculum content so there is less teacher input into it, but they also decrease collaborative decision making in curricular and assessment matters, threatening the kind of work made possible by local control (Hantzopoulos 2004). In a climate of more restrictive curricular and administrative control, schools committed to faculty collaboration like HPA seek political allies that share similar practices. As a result, HPA and other critical small schools have created and become affiliated with local, state, and national networks that support their unique missions and critical pedagogies. This phenomenon, described in greater detail below, reveals how teachers in these schools align with one another to form a collective political force with the potential to influence or reverse regional and/or national policy around issues like standardized high-stakes testing.

Beyond HPA: Growing a Critical Small Schools Movement through Interschool Networks

Since its inception, HPA has viewed the high-stakes testing movement as antithetical to its mission and has been active in the local and national movement against it. In particular, it has worked with the Consortium, a local and state advocacy network of 28 critical small schools in the state that includes teachers, administrators, parents, students, academics, and community activists. When the state commissioner announced the plan to phase in five high-stakes tests as a graduation requisite in 1996, the Consortium was launched. School personnel at HPA have been extremely vocal and influential in this struggle, as the founding principal was co-chair of the Consortium until May 2006.

By co-opting the term "standards" in the group's name, the Consortium challenges the belief that high standards can only be achieved through high-stakes standardized reform and assessment. Instead of these exams, the Consortium schools have devised a system of project-based assessment tasks (PBATs) in which students demonstrate accomplishment in "analytic thinking, reading comprehension, research writing skills, the application of mathematical computation and problem-solving skills, computer technology, the utilization of the scientific method in undertaking science research, appreciation of and performance skills in the arts, service learning and school to career skills" (New York Performance Standards Consortium 2008).[3] Students present this work to a panel of teachers and external evaluators, who review the work and weigh in on the final evaluation using the Consortium designed (and state approved) rubrics.

Advocacy and activism by teachers and other Consortium members are an essential part of its work. In particular, the Consortium has consistently documented how PBATs meet and exceed New York State Regents standards through this system of performance-based tasks. Thus, the purpose of forming this alliance is many-fold; it is not only to help nurture a network of schools in which teachers practice critical pedagogy—including alternative forms of assessment—but also to help reverse the current mandate that all high school students must pass Regents exams to graduate. Moreover, the Center for Inquiry, which is the professional development arm of the Consortium, sponsors year-long workshops for teachers in member schools, providing opportunities for them to share their best practices and to work together on creating their own assessments.

HPA also joined the CES SSP in 2003, a national network of small schools that is committed to personalized learning environments and democratic education. Though CES views school reform as "an inescapably local phenomenon," CES hopes to establish common ground among affiliate schools, described as "groups of people working together, building a shared vision and drawing on the community's strengths, history, and local flavor" (Coalition of Essential Schools 2008).[4] While the CES SSP was specifically designed to create and support equitable, personalized, and high-quality small schools across the country, it was also established to position CES as a major contender in the national conversation of effective school reform, with the hope of redirecting the small schools movement. With more than forty schools active nationally, the SSP meets four times a year in different locations, providing a forum for members to come together and share ideas and practices with one another. One of the major goals is to provide "an opportunity for professional development, allowing participants to learn from each other while also serving as a forum to influence public policy and create a political climate that sustains their work" (Coalition of Essential Schools 2008).

One aspect of the SSP that facilitates this type of collaborative and critical reflection among teachers in various schools is Critical Friends Visits (CFVs). When network meetings take place, schools in the host city open their doors to those in the network who want to visit them. Generally, each visit is structured around an essential question determined by the host school, and teachers, at the end of each visit, have a chance to give feedback on what they saw to the host school community. In 2005, when the SSP meeting was in New York, HPA hosted a visit that asked guests to observe how democratic practices were manifest (or not) in the curriculum and structures that teachers, students, and administrators have jointly designed.

Two years later, in April 2007, HPA sponsored a more extensive CFV for its new mentee school, the James Baldwin School (JBS), outside of the regular network meetings.[5] According to the SSP Guidelines, each new small school must host a visit with the assistance of their mentor school "in the spirit of growth and continuous improvement" (Coalition of Essential Schools 2008). Meant to be a reciprocal learning experience for both the teachers at the host school and the members of the visiting team, visitors are not only supposed to give feedback to the host school but also to look for and learn about best practices to take back to their school. The visiting teams generally consist of mentor school participants—primarily teachers—and three to five other participants from various network schools. This visit, elaborated in the following text, provides an illustrative example of how these networks work as possible vehicles for teacher participation in change efforts beyond their particular classrooms and schools.

Before the visit, teachers at JBS completed a self-assessment on school practices and determined an essential question for the visitors related to a CES common principle or benchmark. After meeting and consulting with the mentor liaison from HPA, the staff at JBS decided to focus on two essential questions for the visit: (1) How is the principle "student as worker, teacher as coach" manifest in the daily practices of school life? (2) What evidence do you see of the goal of personalization of education being applied to all students? The visiting team, which included teachers from New York, Colorado, and California, received this information, along with general information about the school, well before the actual visit with the expectation that they would review these materials before arrival.

The actual visit commenced with an informal breakfast and a warm welcome by the JBS principal, members of the teaching staff, and students. They provided the visitors with an introduction to the school, an overview of the visit, and then paired each teacher with a student ambassador who would serve as a guide for the two days. The participants received reflection sheets to record their notes and a schedule of classes to attend, and they went off on their separate ways with the student ambassadors. At the end of the day, there was a short debriefing session among the participants and the JBS staff and students, mostly pertaining to clarification questions that arose about the school.

The second day was structured differently and began with JBS teachers presenting a dilemma to the Critical Friends participants, facilitated by the HPA teacher who was the mentor liaison. Using a "Consultancy Protocol"[6] to unpack their question and obtain feedback, JBS posed the question, "How can we build and maintain our school culture, emphasizing personalization, community, and democracy, while nearly doubling in size next year?" Because JBS opened in 2005, the school became subject to more stringent space, curricular, and student enrollment requirements, despite being a replication of HPA and intending to open with fewer students.

This exploration of the tension between expansion and maintaining a personalized culture permeated the rest of the visit, as the participating team later interviewed students in focus groups about school culture and growth. Only select students participated in this phase of inquiry with the visitors (because of scheduling), though following the focus groups was a full town meeting in which all members of the JBS community engaged in a discussion on how to deal with the impending school growth. The CFV participants witnessed this, and then they were sent to advisories[7] after the town meeting to debrief about the discussion.

The entire visit culminated in a teacher/student debriefing session with the visitors at the end of the day. Using a protocol called "School Visit Debrief," the visitors went through rounds of observational data to provide feedback to the host school. Participants shed light on what they observed. The visit helped to catalyze JBS staff to think anew on their situation and to brainstorm ways of coming up with potential organic solutions to their dilemma, as well as ways in which they might bolster their practices of "student as worker, teacher as coach." According to one JBS teacher, "We gained so much from the Critical Friends Visit: feedback on the importance of more discussion in classroom, more common practices, involving students as leaders in school more explicitly" (Interview, May 2, 2007). Another stated, "The clear and consistent feedback was especially useful in affirming the importance of professional development for new teachers ... [realizing] that we should commit to this more" (Interview, May 2, 2007). In addition, the visitors, including the teachers from HPA, commented that they had learned many things from the visit to take back to their own school settings, thus demonstrating the sharing of ideas around pedagogy and school structure that is cultivated in these networks.

Although the description of the CFV makes evident how schools might work in partnership to support and learn from each others' distinct practices, there are also important policy implications that can potentially arise out of such collaborations. In fact, when HPA decided to replicate itself and open JBS, the staff intentionally sought CES financial support over other sources because of its philosophy of school design and its commitment to reenergizing the small schools movement. As one participant from the CFV stated:

> I think that the empowerment that one feels in being part of a conversation with educators of like values, but with diverse perspectives and experiences, who have come together from good or great schools across the country ... this empowerment is what makes tackling wider reform possible. You feel stronger and part of a wider community of dedicated educators. You feel then up to the task of taking on state and national and the even often

daunting scope of local reform...I always feel empowered in this way after meeting with the CES Small Schools Network. (Interview, May 1, 2007)

Though these networks are not necessarily a remedy for standardization, these words indicate how the network might provide a local, organic, and partial antidote to centralized educational reform at the state and national levels. The JBS example above also exemplifies one of the main purposes of the SSP, which is to "provide an opportunity for professional development, allowing participants to learn from each other while also serving as a forum to influence public policy and create a political climate that sustains their work" (Coalition of Essential Schools 2008).

As Michael and Susan Klonsky (2008) suggest, one of "the most effective response[s] to top-down reform is for teachers, parents and communities to act together powerfully. Real reform is not confined just to the policy arena but, more often, is generated by the way in which educators engage one another and the students they are teaching" (163). Thus, these networks have potential to go beyond strengthening best practices within schools and provide space for cross-fertilization, collaboration, and reflection *among* teachers that can also lead to ways of confronting mandates (in this last circumstance, forced school expansion from the Department of Education), finding local solutions, and possibly reversing policy. In the case of the Consortium, this strategy has proven quite effective. As of June 24, 2008, the New York State Board of Regents voted to extend the Consortium's variance—meaning that Consortium schools can continue to administer their PBATs in lieu of the other Regents exams—for another five years with only the English Language Arts Regents exam required for graduation. This is a remarkable victory given that three years ago, the same board stated that the Consortium schools would eventually have to give all of the Regents exams in addition to their PBATs.

The Possibilities and Limitations of Teacher Participation in a Critical Small Schools Movement

Through banding with both of these professional networks, HPA teachers meet with like-minded colleagues to collaborate, address dilemmas, and contemplate local solutions to them. Although the schools in the networks were conceived differently, interschool organizations allow teachers at HPA and the other member schools to share and debate some fundamental beliefs about education, augmenting each teacher's own practice and, collectively, the practices in their schools. Perhaps what is most compelling about these relationships is that they serve as potential ways to safeguard the local forms of critical education being practiced by teachers in each school through the creation of broader regional and national collectives. By associating with larger networks, the critical, democratic practices and distinctive school structures in these small schools are shielded in an oppositional climate of neoliberal educational reform. In many ways, each individual school embraces the mantra "think locally, act locally"[8] as a form of social activism and preservation because school activists operate in the interests of preserving their exclusive school cultures by uniting with other reformers who wish to do the same (see Esteva and Prakash 1998).

Moreover, the schools in these collectives can help shape and transform the broader education agenda to inform, strengthen, and revitalize a national critical public small schools movement. For instance, when the SSP was launched, CES had not officially issued a statement against high-stakes testing (though their guiding principles favored project and inquiry-based assessment). As a direct result of the conversations that various teachers in the network were having around this issue, CES decided to take a stand. This declaration by CES reflects an interesting twist in the relationships between local school actors such as teachers and the broader networks they build, demonstrating that through these alliances, teachers in critical small schools have the potential to shape larger forces at the regional, national, and even international levels. In fact, schools from the Netherlands and Australia have visited HPA to observe their practices; their interest in HPA was generated through its affiliation with CES.

To move the critical SSP forward, there needs to be conscious self-reflection about the ways in which broader networks can reshape local schools and the practices of teachers in them. Because both the Consortium and the CES network offer reflective professional development that encourage schools to be critical of their own practices, a worthwhile endeavor would be to document how teachers and schools have changed as a result of these affiliations. In addition, as Taylor and Wilkinson have elucidated in chapters 4 and 5, respectively, participatory democracy has its limitations and does not always deliver what it promises. For instance, it is possible that some teachers do not participate as fully (or at all) as others in democratic school practices, as seen in the situation with the parent-councilors in Wilkinson's study of participatory school governance in Brazil. Moreover, some administrators in the critical small school movement may only permit "conditional" critical pedagogy, something akin to the *prix fix* menu of participatory practices described by Taylor in her study of Tanzania. There may also be cases where teachers and students' involvement in school governance is in fact limited to principal-designated activities and spaces, thereby setting conditions on the extent to which ground-up school reforms are actually enacted. Further research at critical small schools, especially with the teachers shaping the investigations, would expand our understanding of the potential and the limitations on the movement. In this way, it will be possible to further engage in and grapple with the challenges of moving the project of small schools and democratic participation in education in a new direction.

Notes

1. In New York City, small schools have generally less than 400 students. Mayor Michael Bloomberg and Education Chancellor Joel Klein have touted small schools as the priority of their educational (and political) agendas and have offered them significant support (Gootman 2006a). More than 200 new small schools have opened in the city; at least 100 of these have been launched since 2006.
2. Some examples are designing computer programs, examining environmental racism, or investigating the root causes of slavery in the United States (see New York Performance Standards Consortium 2008).
3. For details see the New York Performance Standards Consortium Web site: http://www.performanceassessment.org.
4. For further information see the Coalition of Essential Schools Web site: http://www.essentialschools.org.
5. Through CES, HPA was awarded a replication grant to launch the James Baldwin School in the fall of 2005.
6. See National School Reform Faculty (www.nsrfharmony.org) for information on Critical Friends protocols.
7. Advisories are small student-centered classes at HPA in which the basic curriculum consists of student-generated topics.
8. This mantra was coined by Esteva and Prakash (1998) to describe local collectives that counter the global orientation in international development discourse. Escobar (2000) also describes its meaning as "localities as locally connecting with each other in terms of solidarity, in order to resist the global thinking that dominates imperialist and neo-liberal approaches, because any type of global thinking, in sum, might may be too complacent with the same reconstruction of the world" (172).

Part 3

Examining the Political Economy of Diversity

Chapter 7

"Migration Nation"
Intercultural Education and Anti-Racism as Symbolic Violence in Celtic Tiger Ireland

Audrey Bryan

Cultural diversity has historically been a feature of Irish society, as evidenced by the presence of a host of minority groups, including Travellers, blacks (often referred to as black-Irish), Jews, and Asians.[1] Despite this diversity, the "Celtic Tiger" era is often popularly (if erroneously) understood to signal Ireland's transition from a monocultural to a multicultural society. This chapter examines the interplay of inter/national forces that influenced the evolution of educational policy responses to cultural diversity and the intensification of racism in the Republic of Ireland during the so-called Celtic Tiger era. The idea of the Celtic Tiger economy, akin to the "tiger" economies of south-east Asia, first emerged in the mid-1990s, as evidence of Ireland's nascent economic boom began to accumulate (Coulter 2003). This period of unprecedented economic growth was associated with changes in the operation of global capitalism and the convenient base that the Republic offered to multinational corporations seeking to expand operations in Europe (e.g., Coulter 2003; Garner 2004). The accompanying immigration trend, which coincided with European Union (EU) enlargement, resulted in a newfound emphasis on interculturalism and anti-racism at multiple levels of Irish society, including schools.

As Valdiviezo demonstrates in the next chapter, the discourse of interculturalism has become increasingly transnational in scope, and it is related to a global educational policy agenda regarding the provision of culturally and/or linguistically sensitive education. The recent focus on interculturalism and anti-racism in an Irish context is in part a response to broader inter/national influences and occurrences, including support for EU member states to engage in national-level initiatives designed to foster "intercultural dialogue" in light of the growing diversity within the EU. It can also be seen as a response to the intensification of racism in Irish society during the Celtic Tiger era, evidenced by, *inter alia,* a wave of hostility toward refugees and asylum seekers (fueled by certain media and political representatives), and the establishment in 1998 of an anti-immigration lobby known as the Immigration Control Platform (Allen 1999).

In this chapter, I examine how inter/national intercultural policies are enacted, nationally and locally, by combining discourse analysis of national intercultural and anti-racism policy documents with interview and observational data derived from a 12-month critical ethnography of one school's efforts to promote "positive interculturalism." This chapter seeks to highlight the ways in which current configurations of racism in Irish society, as well as official policy responses to the problem, are linked to broader political and economic structures and forces at national and international levels. Like the other chapters in this section, I am committed to examining the political economy of diversity: That is, I consider state-crafted as well as locally created strategies for "managing" diversity as they are shaped by the confluence of local and global economic and political forces. More specifically, I outline a range of political-economic arrangements that were implicated in the intensification of racism during the Celtic Tiger era, which resulted in a dramatic increase in wealth inequality and poverty, and in the proliferation of low-paid, insecure, and part-time employment alongside dramatic increases in corporate profits (Loyal 2003). I then seek to demonstrate the extent to which structural explanations of this nature, which emphasize the political-economic conditions that help to produce racism in the first instance, are typically ignored or underplayed in recently produced policy documents and guidelines that are purported to help alleviate racism in Irish society. This tendency, I argue, has important implications for how young people understand racism in an Irish context.

Conceptual Framework: Interculturalism and Anti-racism as Symbolic Violence

This chapter contributes to studies that underscore interculturalism's "susceptibility to appropriation" by the state and dominant groups, as we have seen in such contexts as Australia, Canada, the United States, and the United Kingdom (McCarthy, Crichlow, Dimitriadis, and Dolby 2005, xxiii). Drawing on the intellectual oeuvre of Pierre Bourdieu, I characterize interculturalism as a form of "symbolic violence" to convey how racial domination operates on an intimate and subtle level within the context of social and educational policies, practices, and curricula that are purportedly liberal and multiculturalist (Bourdieu 2001). Symbolic violence describes a mode of domination that is exercised on individuals in a subtle and symbolic (as opposed to physical) manner, through such channels as communication and cognition (ibid.).

Before presenting some of the study's major findings, which suggest that racial inequality is more likely to be reproduced through policies and practices intended to be egalitarian and anti-racist, I provide a brief overview of the birth of the Celtic Tiger economy as the contextual backdrop for the evolution of interculturalism and anti-racism in light of intensified racism in Ireland.

The Birth of the Celtic Tiger Era and the Evolution of Interculturalism

The idea that Irish society is multicultural is commonly understood as a recent development in Ireland's demographic history, despite the historical presence of culturally diverse groups in Ireland. This fact contrasts sharply with other geographical contexts, such as North America, where the notion of multiculturalism is entrenched in national consciousness and where the nation is imagined as having a long history rooted in multicultural beginnings (Montgomery 2005). The perception of Irish cultural homogeneity has been linked to the political project of Irish nationalism that emerged following Ireland's independence from Britain in the 1920s. This project, which was based upon racialized and exclusionary foundations, constructed Irishness as a homogenous entity that was essentially white, Catholic, nationalistic, and rural (see e.g., McVeigh and Lentin 2002; Loyal 2003; Connolly 2006).

In the 1980s, Ireland experienced a severe economic recession, characterized by high unemployment rates, substantial public debt, and mass emigration. It lagged so far behind most other EU member states on all indices of economic performance that it bore many of the hallmarks of a "Third World" country (Coulter 2003). Yet, less than a decade later, many politicians and social commentators were celebrating an economic boom that earned the Irish economy the label the "Celtic Tiger." Fiscal and other incentives (including very low taxation on profits) made Ireland an investment paradise for multinational firms seeking to gain access to the EU market, especially those involved in the information technology and pharmaceuticals industries, resulting in a major increase in foreign direct investment. The unemployment rate fell drastically: from more than 15 percent in 1993, to 6 percent in 1999, to 4.2 percent in January of 2006 (EUROSTAT 2006). By the end of the 1990s, economic experts were warning that a labor shortage could pose a serious problem to continued economic growth. In an effort to foster greater economic growth and to meet demands for labor, the government reached out to so-called non-Irish nationals and returning Irish emigrants alike. Simultaneously, social unrest and poverty in various parts of the world were forcing a small yet significant number of refugees and asylum seekers, primarily from African and Eastern European states, to seek refuge in Ireland. According to the 2006 census, 15 percent of those normally resident in Ireland were born outside the state, and 10 percent were of foreign nationality (Central Statistics Office 2007).

As the population of Ireland became increasingly more ethnically diverse in the late 1990s and first decade of the twenty-first century, evidence of growing anti-immigrant sentiment became apparent, exemplified by sensationalist media reports that depicted immigrants, refugees, and asylum seekers in a predominantly negative and stereotypical light (Devereux and Breen 2004; Devereux, Breen, and Haynes 2005, 2006a, 2006b). Eurobarometer polls designed to measure levels of racism and xenophobia in European member states were carried out in 1997 and 2000 and provide further evidence of rising levels of hostility toward minority groups in an Irish context. Whereas in 1997, 16 percent of Irish respondents agreed that the presence of people from minority groups offered grounds for insecurity, three years later, the percentage of those who agreed with that view had increased to 42 percent (Thalhammer, Zucha, Enzenhofer, Salfinger, and Ogris 2001). Moreover, the proportion agreeing that it is a good thing for society to be made up of people from different races, religions, and cultures fell from 76 percent in 1997 to 61 percent in 2000 (ibid.). Recent surveys of experiences and perceptions of racism among minority groups are also indicative of levels of racial hostility in an Irish context. A 2005 survey of discrimination against minorities in Ireland conducted by the Economic and Social Council (ESRI) found that 35 percent of recent migrants had experienced harassment on the street, on public transport, or in public places, with more than half of all "black South/Central Africans" having experienced this form of discrimination. In addition, a fifth of the sample who were entitled to work reported having experienced discrimination in gaining access to employment (McGinnity, O'Connell, Quinn, and Williams 2006).

Within this context of an increasingly ethnically diverse population and the emergence of "new configurations" of racism (Garner 2004, 228), Irish social and educational policy began to reflect a commitment to interculturalism and anti-racism (Lodge and Lynch 2004). However, as Valdiviezo highlights, the development and implementation of state intercultural policies are often also the result of prevailing global agendas and efforts to demonstrate national commitment to equality, justice, and human rights, part of an effort to garner respect from the wider international community and to obtain additional external financial support for educational and social projects. The rationale for developing anti-racism and intercultural policies in Ireland is similarly driven by inter/national and economic considerations. For example, national anti-racism policy documents refer directly to the need to preserve the nation's "international image" as a warm and welcoming place to visit and to live, and to its "international reputation built on proactively supporting human rights and speaking out on human rights abuses at a global level" (Department of Justice, Equality and Law Reform 2005, 41). The threat that racism poses to the Irish economy in terms of lost revenue from tourism and the need to develop intercultural workplaces to ensure competitiveness within a global economy also feature as rationales for intercultural and anti-racism strategies (ibid.). The following section provides a more detailed overview of these national-level policy responses, in the broader social and educational domains.

Planning for and Managing Diversity

Ireland's first ever *National Action Plan Against Racism* (NPAR) was officially launched by the *Taoiseach* (Prime Minister) and the Minister for Justice in January of 2005. NPAR arose from a commitment made by the Irish government at the UN World Conference against Racism in 2001. As a framework for public policymaking, the plan promotes interculturalism as an effective means by which racial discrimination can be opposed and ultimately eradicated. NPAR is the most comprehensive articulation of official thinking on interculturalism in Ireland; as the cornerstone of the government's anti-racism policy, its overall aim is to "provide strategic direction to combat racism and to develop a more inclusive, intercultural society in Ireland based on a commitment to inclusion by design, not as an add-on or afterthought, and based on policies that promote interaction, equality of opportunity, understanding and respect" (Department of Justice, Equality and Law Reform 2005, 27). In July of 2007, the Office of the Minister for Integration (OMI) was established "as a response to the recognition of the scale of migration to Ireland in the last decade or so, particularly since 2004" (Office of the Minister for Integration 2008, 67). As the name suggests, this office has responsibility for developing and implementing a national integration strategy, a major focus of which is on "diversity management" or "properly managed immigration" (8). The strategy portrays integration as a two-way process requiring mutual adaptation on the part of "host" and "new" communities. As part of its remit, the OMI promotes national campaigns aimed at challenging racism and fostering understanding of diversity, including an annual "Anti-Racist Workplace Week" and the "measurement of integration outcomes" (OMI 2008, 24).

Educational Responses

In line with the OMI's perception that "efforts in education are critical to preparing immigrants, and particularly their descendants, to be more successful and more active participants in society," the Department of Education and Science (DES) is one of three government departments designated as having a central role to play in "dealing with the integration of migrants into Irish society" (OMI 2008, 67). A dedicated "integration unit" within DES was established in October of 2007, and the DES recently expressed a commitment to develop a National Intercultural Education Strategy. Since the late 1990s, the DES has made provision for the appointment of "language support teachers" in schools to provide nonnative English speakers with sufficient verbal and written skills to facilitate their integration into a predominantly English "mainstream curriculum in as short a time as possible" (Joint Committee on Education and Science 2004, 14). Students are entitled to two hours of language support tuition per week for a maximum of two years (Nowlan 2008), which typically involves the withdrawal of pupils (individually or in groups) from mainstream classes (Nowlan 2008; for critiques of this program, see Ward 2004; Devine 2005; McDaid 2007).

A host of intercultural educational materials and guidelines have been produced by various statutory and nonstatutory agencies in recent years. Most notably, the National Council for Curriculum and Assessment (NCCA), an advisory body to the DES with responsibility for developing and assessing school curricula, published intercultural guidelines for both primary and secondary schools, which focus on "mediat[ing] and adapt[ing] the existing curricula to reflect the emergence of a more culturally diverse society in Ireland" (Department of Justice, Equality and Law Reform 2005, 110). Whereas earlier educational policy documents have been critiqued for their failure to devote "substantive treatment" to the "issue" of diversity (DES 2002, 15), the NCCA guidelines promote intercultural education as a means of underscoring "the normality of diversity in all parts of human life" (NCCA 2006, i). The notion of respecting, celebrating, and recognizing diversity as normal is identified as one of two "core focal points" of intercultural education, alongside the promotion of "equality and human rights, challeng[ing] unfair discrimination, and promot[ing] the values upon which equality is built" (NCCA 2005, 3). Intercultural education is believed to "help prevent racism" by enabling students to "develop positive emotional responses to diversity and an empathy with those discriminated against" as well as enabling them to "recognise and challenge discrimination and prejudice" where they exist (NCCA 2005, 21). As such, intercultural education is deemed "one of the key responses to the changing shape of Irish society and to the existence of racism and discriminatory attitudes in Ireland" (NCCA 2005, 17).

The remainder of this chapter considers how national anti-racism and intercultural policies are instantiated and interpreted locally by profiling one school's efforts to promote a policy of "positive interculturalism" in the greater Dublin area. I argue that the potential for intercultural education to "help alleviate racism," as its goals suggest, is blunted because its implementation coincided with a set of political-economic arrangements that provided the structural basis for the intensification of racism in Celtic Tiger Ireland, including accelerated income and wealth polarization, soaring cost of living (exacerbated by an out-of-control property market), new modes of employment, and increasingly flexible labor market practices (in the form of badly paid, poorly protected, part-time, and temporary employment contracts) (Loyal 2003; Garner 2004). From this vantage point, racism can also be understood as a response to socioeconomic conditions that heighten material vulnerabilities and anxieties—anxieties which are projected onto vulnerable groups such as Travellers, asylum seekers, and economic migrants who are deemed privileged recipients of diminishing national resources, such as welfare payments, jobs, or land (Rizvi 1991; Hage 2003; Garner 2004). As I argue in more detail below, at both national and local levels, these structural dimensions of racism are eclipsed by individualized and depoliticized explanations that attribute racism to individual ignorance, a lack of appreciation or awareness of other cultures, or not being used to "difference."

Methodology

The study combines critical discourse analysis (CDA) of national intercultural and anti-racism policy documents with interview and observational data derived from a 12-month critical ethnography of an ethnically diverse school that has embraced interculturalism, to examine the effects of national-level intercultural policy and discourse at the local level. Adopting a vertical case study approach, I examined interculturalism as it is conceived and enacted at multiple levels and in multiple arenas in Ireland (national-level policy; curriculum materials; microlevel events in school), and I located this analysis within a broader consideration of the macrostructural context of racism during the Celtic Tiger era. The interviews and observations enabled a "thick" understanding of meso- and microlevel racialized dynamics and experiences, as well as understandings of intercultural concepts and practices, whereas the CDA focused on the production of meaning of key concepts related to interculturalism by key players within the national political and educational arenas (e.g., Fairclough 1995, 2003; Levett Kottler, Burman, and Parker 1997; Van Dijk 1997). This critical analysis of national anti-racism policy documents, textbooks, and other instructional materials designed for use with lower secondary school students attending school in the Irish Republic involved a multilayered process of reading, writing, and interpreting each of the texts to derive recurring patterns and themes. I examined various degrees of presence or absence in the texts, such as "foreground information" (those ideas that are present and emphasized), "background information" (those ideas that are explicitly mentioned but de-emphasized), "presupposed information" (that information which is present at the level of implied or suggested meaning) and "absent information" (Fairclough 1995).

I examined intercultural policy and practice as it is enacted "on the ground" at a coeducational secondary school that I call Blossom Hill College (BHC) (a pseudonym), located in a middle-class suburb in the greater Dublin area. Adopting a policy of "positive interculturalism," BHC has been identified as a model of "best practice" in promoting inclusivity, interculturalism, and equality, and approximately 10 percent of its student body is "international."[2] Between September 2004 and December 2005, I observed classroom lessons and school events and conducted more than 30 interviews with students and 5 interviews with teachers.

National Discourses on Diversity: "Migrants with a Contribution to Make"

The discourse on diversity in Ireland that is evoked in national social policy documents and curriculum guidelines is ostensibly one of "celebratory interculturalism," which recognizes and welcomes the fact that Ireland is now a more diverse society in which "people of different cultural and ethnic backgrounds enrich society (Parker 2001, 74). Critics of state-sanctioned multiculturalism more broadly have critiqued this *raison d'être* of the intercultural project on the grounds that the discourse of "respecting," "celebrating," "valuing," and "appreciating" diversity has the effect of denying the possibility of a national "we," which is itself diverse, and ultimately entrenching power relations between culturally dominant and minority groupings in society by positioning culturally dominant groups at the center of the national cultural map around which "otherness" is located (see Ang 1996; Hage 1998; Bryan 2008). Hage (1998) argues that the discourse of cultural enrichment assigns to minority cultures "a different mode of existence," defining their worth in terms of their function as enriching dominant cultures and hence their existence primarily, if not exclusively, in terms of the benefits they offer for dominant groups (121). Drawing on Hage's analysis, I maintain that state-sanctioned interculturalism positions culturally dominant groupings in Irish society as the "tolerater" or "embracer" of difference, who get to decree the acceptability (or otherwise) of the ethnic Other. From this vantage point, the very expression of acceptance implies that such acceptance in Irish society is conditional and could be withdrawn were migrants to be deemed undeserving of this reception (ibid.).

Recent "integration" strategy statements highlight the conditional and contingent nature of Irish state-sanctioned interculturalism. The gains to be had from migration are clear in the government's commitment to "immigration laws that control and facilitate access to Ireland for skilled migrants *with a contribution to make*" (OMI 2008, 9, emphasis added) and in reference to the "societal gains from properly managed immigration" (OMI 2008, 8). Statements such as these, which bestow acceptance exclusively on those minorities who are perceived to be skilled, hardworking, and a benefit to Irish society in economic terms, have the effect of entrenching power relations between the acceptor and those whom they accept, and of legitimizing negative responses toward those who are deemed less skilled, and, thus, who do not have a contribution to make, as defined by dominant cultural groups. From this perspective, negative reaction to unskilled or "undeserving" minorities who do not make the kinds of economic and social contributions deemed healthy for Ireland becomes natural, acceptable, or at least understandable (Gillborn 1995; Blommaert and Verschueren 1998). State-level "anti-racist" discourses of this nature, which explicitly or implicitly suggest that immigrants are welcome so long as they benefit Irish society and are therefore deserving of its resources, serve as a basis for promoting—as opposed to contesting—racism against those who are deemed undeserving of the state's "generosity."

Getting to Know Racism?

The prevailing view of racism that is presented in national policy documents positions racism as an individual, psychological phenomenon,
divorced from political-economic considerations or arrangements that provide the structural reasons for racism in the first instance. Although there
are occasional references to institutional racism in documents like NPAR,
its "unintentional" or "inadvertent" aspects are typically privileged, as are
individualistic explanations, such as a "lack of thought" and "understanding," as the basis for institutional racism. Furthermore, the state's role in
perpetuating institutional racism is discussed solely at the level of its failure
to provide adequate services to ethnic minorities.

Apart from these brief references to institutional racism, most representations of racism in NPAR and related documents perpetuate the "rotten
apple theory" (Henriques 1984). This theory maintains that there are racists in society but that there is nothing inherently or institutionally racist
about the structure of society (Troyna 1993). Focusing on the few "rotten
apples," it has the effect of deflecting attention from the systemic features
of racism and prioritizing ineffective, and at times counterproductive,
"softly-softly," or gentle approaches to anti-racism campaigns and programs that do not address the root (i.e., structural or institutional) causes
of racism (Bonnett 1993). Evidence of this individualistic understanding of
racism is provided in the foreword to the Final Report on Activities of the
Government-Sponsored National Anti-Racism Awareness Program (published as part of NPAR) by the Chairperson of the Campaign's Steering
Committee, Joe McDonagh:

> As I have stated on many occasions in the past, Irish society is now a mul-
> ticultural society. We must accept the responsibilities and challenges that
> change brings to us. Irish people are traditionally generous, friendly and
> hospitable. It would be wrong to allow fear of strangers and intolerance to
> spoil this traditional spirit and change our attitudes towards the minority
> ethnic people who are part of Irish society. (Know Racism 2005, 6)

These statements present an idealized and stereotypical notion of Irish people as "traditionally generous, friendly and hospitable," while simultaneously
positing minorities as "strangers." This has the effect of working collectively
to prevent "spoiling" the presumably shared elements of "traditional" Irish
society (Gillborn 1995; Blommaert and Verschueren 1998). In other words,
"fear" and "intolerance" are presented as unfortunate tendencies—not necessarily racist responses—and diversity as a new phenomenon that should
be embraced because the Irish have "traditionally" welcomed "strangers."

Further, NPAR and related documents constantly emphasize "combating," "eradicating," "stopping," and "ending" racism. This strategy problematically reinforces the idea of anti-racism as some kind of definitive end point rather than as an ongoing commitment to social justice that must be continuously reaffirmed through structural and institutional change (Montgomery 2005). As implied in the title of the Know Racism Campaign, this initiative is premised upon the notion that racism is something that can be tackled, simply by saying "No" to racism. It assumes that by getting to "Know" racism" by "increasing our knowledge and awareness of racism," through increased "interaction, understanding and integration," and by "promoting an appreciation of cultural diversity" (Know Racism 2005, 3), racism itself will disappear.

This approach is not surprising given that the campaign positions racism as an individual or interpersonal manifestation rather than as a multidimensional phenomenon that manifests itself at multiple levels and is inscribed in broader social processes and practices. In other words, the anti-racism awareness campaign conceives of racism largely within a prejudice/discrimination framework that reduces racism to irrational and individualized problems of thought (e.g., a "sense of superiority," "fear or insecurity," and/or a "lack of knowledge") and behavior (such as "extreme" acts like "racially motivated abuse, direct discrimination and violence"). The emphasis on "interaction, understanding and integration" conveys the extent to which these efforts are based in a Contact Hypothesis framework, which proposes that contact between members of different groups will result in a reduction of prejudice between these groups and an increase in positive and tolerant attitudes (Allport 1954). While interpersonal relations may have the potential to reduce racist attitudes and behaviors, in the absence of any consideration of the broader social processes, institutions, and structures that help to create and sustain racial oppression, this hypothesis restricts the nature and causes of racism to the realm of individual ignorance and misunderstanding that can be combated through greater "contact" with the Other.

From this perspective, policies that operate according to contact theory have the effect of enabling governments to distance themselves from racist incidents, thereby casting the state in a redemptive, nonracist, or anti-racist light. Campaigns like Know Racism that present racism as an individual pathology render any examination of the economic and social conditions that help produce racism unnecessary (Rizvi 1991). The kinds of anti-racism programs that the state is likely to sanction, then, are based on a "soft-antiracism" model, which occludes consideration of the systemic nature of racism in society and of the need to disrupt the status quo.

Having provided a critical overview of interculturalism and anti-racism as they are conceived at the national level, I now seek to examine how interculturalism is conceived and practiced at the local school level, drawing on a 12-month ethnography of BHC.

"Positive Interculturalism" at Blossom Hill College

Similar to the celebratory tone of national-level intercultural and anti-racism policy, BHC is committed to the idea of interculturalism, and it has adopted a policy of "positive interculturalism," incorporating a range of activities that celebrate the cultural diversity of its student body. Various symbolic gestures to promote positive representations of diversity and to define the BHC community as multicultural are evident throughout the school. For example, there are posters displayed throughout the school that read "diversity is a strength, not a weakness" and "BHC cherish[es] the students of all nations equally." There is also a multilingual "welcome" sign and flags representing all the different nationalities represented at the school, which are proudly displayed in the main entrance. This pride in the multicultural composition of the school is also evident in official school publications, such as the school yearbook, and the virtues of cultural diversity are extolled, both rhetorically and symbolically, at major school events, including the school's anniversary celebration and graduation ceremonies, and at interfaith ceremonies.

Facilitating and ensuring positive interaction and social integration between "national students and international students" at BHC is a major aim of the school's intercultural strategy. In one of my initial conversations with the vice principal, Miss Deagan (a pseudonym), I was informed that the school had conducted a survey of its "international" students, which suggested that there was "not a lot of mixing" between "international" students and their "Irish peers." The belief that increased contact can facilitate reduced racial antagonism has also influenced intercultural practices at BHC, such as the development of a "buddy system" linking national and international students, and the organization of social evenings, such as an "Afro-Caribbean" evening for parents. As Miss Jones, the teacher with responsibility for coordinating language support and interculturalism at BHC explained: "My post [is] actually...totally taken up with devising ways of providing opportunities for national students and international students to come together" (Interview, September 22, 2005).

Hage (1998) maintains that concerns over minority groups not integrating sufficiently and the associated desire to facilitate integration is, in effect, a desire for more "supervised integration" on the part of the culturally dominant group, which grants the dominant culture the ability to manage diversity (239). Paradoxically, these supervised integration activities at BHC coexist alongside other intercultural practices that actively segregate (albeit temporarily) linguistic minority students from their peers. For example, the practice of withdrawing linguistic minority students at BHC from mainstream classes for language support can compound students' preexisting perceptions of being made to feel like the Other by drawing attention to their "difference" in front of teachers and peers (Devine 2005). Some of the minority students with whom I spoke at BHC were quick to point out the marginalizing and Othering effects of the model of language support offered at BHC. Shoma, for example, explained the pressure she felt to "act like" her "Irish" peers, to counteract the sense of being perceived as Other, or different, which she felt as a consequence of having to do "extra English" classes:

> And they think that you are…they already think that you are different, when you are doing extra English (language support) and doing fasting and you are doing different, so they treat you more different so you have to try and act like them so they can think "well they are actually normal." But otherwise they would be thinking "why are you so different?" You have to fast, and you have to pray five times a day, you have to do this and that, you have to go to extra English. (Interview, April 11, 2005)

In addition to the language support structure, some of BHC's intercultural gestures had clear, albeit unintentional, negative consequences for at least some minority students. For instance, one such effort to sensitize members of the school community to the cultural or religious practices of ethnic minorities that proved unintentionally detrimental was the placing of signs asking students to be "mindful" during the month of Ramadan. Instead of promoting greater awareness, or understanding of the Islamic faith, this call for "mindfulness" resulted in taunting and teasing of some fasting students by their non-Muslim peers. As one student explained:

> Some children what they do is they start making fun of you and stuff. Like the Ramadan posters, when they went up, people were just making fun of them and stuff and then they used to come up to you and asking you are you fasting and stuff like that. Even when they know stuff about it. (Interview, March 14, 2005)

The logic of intercultural education maintains that "by exposing all children to the social and cultural customs of ethnic minority communities, they will have a greater understanding and tolerance of people from

different backgrounds" (NCCA 2005, 169). This is clearly what the staff
at BHC intended with their program. Yet, the example reveals, contrary
to their intended aim of enhancing intercultural understanding and empathy, token gestures like posters can prove detrimental to the extent that
they reinforce minority students' sense of otherness and expose their
cultural practices to ridicule. Indeed, for most minority students with
whom I spoke, racism was a real part of their everyday existence at BHC.
Racist incidents typically took the form of racialized name-calling, verbal
insults, and "slagging" (teasing, mocking, or insulting in a critical manner) although, as the following vignettes reveals, more overt and physical
manifestations of racism were also evident:

> Oh yeah, like they throw stuff at you like in the courtyard. They just like,
> at all the foreign people, they just get their lunch and they start throwing it
> at you. (Interview, September 26, 2005)
>
> Amel was crying and her eye was like swollen. So they had hit her and her
> glasses had broke so it was obvious that something had happened. And
> we pointed out the boys…and I got hit in the head afterwards which
> really hurt, like there was a big red mark over there (pointing to her head).
> (Interview, September 26, 2005)

Despite the physical nature of these attacks, there was still a perception
among some teachers that minority students themselves were often responsible for the racist violence or insults they experienced. As one teacher put it:

> Some Africans…Nigerians, mostly, would tend to be a bit loud and boister-
> ous and to bring it [negative responses] on themselves.…They are inclined
> to strike out, to use the fist, whereas *ours* wouldn't be as quick to do that.
> (Interview, February 7, 2005)

Discourses of this nature, which construct Africans as boisterous, physical, aggressive, and volatile, in opposition to a superior, national, more
pacifist "we," create a tendency for some African students to be labeled as
such. Therefore, they are more likely to be reprimanded and disciplined
by teachers, thereby reinforcing their sense of alienation from school
(Connolly 1998).

The disproportionate disciplining of African students, in particular, at
BHC was exemplified with one of the students with whom I worked closely
during my time at BHC in my capacity as a Language Support Volunteer
tutor at the school. Yvette was a student from the Democratic Republic of
Congo who developed a reputation as a "troublemaker" from her earliest days
at BHC, based on her perceived refusal to make an effort to speak and learn
English, her disruptive behavior, and her apparently disrespectful attitude. In

addition to concerns about her failure to learn English and her insistence on speaking in "some African dialect" with compatriots, Yvette was often reprimanded for her inappropriate "body language," her "refusal to make eye contact," and her tendency to "slouch in the chair." As the year went on, Yvette did, in fact, seem to become increasingly oppositional in her school and classroom behavior, and toward the end of the year, she was suspended from school for initiating a physical fight with her peers. On more than one occasion, I witnessed the Language Support coordinator threatening to withdraw Yvette from language support classes if her behavior and level of English proficiency did not improve. Moreover, this particular administrator frequently told me that I needed to be "tough" on Yvette and to discipline her, something that was not asked of me in relation to students from other backgrounds.

Students' Understandings of Racism at BHC

Consistent with some of the teachers' views about racism at BHC, NPAR, and the school's policy of positive interculturalism, many students I spoke with and/or observed at BHC tended to ascribe the existence of racism in society to individual ignorance, a lack of appreciation or awareness of other cultures, or "not being used" to "difference." The views of the students below typify this view of racism as largely an individual, apolitical problem:

> Yeah, like people only started coming in to Ireland as immigrants in the last [decade] since the Celtic Tiger. And Britain and America have had them for decades, you know. And a lot of people are coming in, so immigrants are blended into their society already well long before us so, we are still getting used to that. (Interview, November 7, 2005)

> I think people feel inferior to people from other countries just because they are not used to them, so maybe they might like act out on them because they are different. (Interview, November 7, 2005)

Related to these students' accounts, which stress lack of awareness and understanding of "others" as the source of racism, some students advocated solutions to racism that were highly reflective of the kinds of solutions advocated in state-sanctioned anti-racist educational campaigns in discussing what can be done to "stop racism":

> It's about teaching people what is right and what is wrong. You've got to teach people that racism is wrong and then they won't be racist. (Interview, November 7, 2005)

> I think nowadays you learn...you kind of pick up on things. Like you learn that being racist is wrong. It's just like people probably learned back

then that being racist is ok so that's what they think is right 'coz they learned...they were told that, told that was right you know. So we are told that's wrong now. (Interview, November 7, 2005)

In general, BHC students' understandings of racism and anti-racism point to the broader curricular, institutional, and societal silences about the systemic nature of racism, silences that cannot be blamed on students themselves because they mirror the situation at the national level (Roman and Stanley 1997). Like the national policies discussed earlier, BHC's efforts to tackle racial, or indeed other forms of inequality and discrimination within a positive intercultural framework, precludes consideration of the more systemic dimensions of these forms of oppression, which contribute to their production in the first place. In this sense, the emphasis on celebrating diversity and promoting interaction has the effect of masking racism, rather than directly working against it. Positive interculturalism thus serves a useful impression management function for the school, casting it as an essentially benign peacemaker through its various efforts to encourage greater contact and positive attitudes between various racialized and ethicized groups (Fitzduff 1995, cited in Connolly 2000).

Reconfiguring Intercultural Education and Anti-Racism

The foregoing analysis suggests that the intercultural and anti-racist policies adopted and/or sanctioned by the Irish government offer an inadequate basis for the realization of a post-racist society because they fail to critique the broader politico-economic and social structures that gave rise to the intensification of racism in Celtic Tiger Ireland in the first instance. To the extent that racism is acknowledged, it is generally reduced to the level of the individual, obfuscating the state's role in reproducing the very thing it purports to eliminate through campaigns and educational initiatives.

The foregoing critique is not meant to imply that schooling has no role in teaching about cultural diversity and against racism. After all, if we are to better understand and overcome racism in society, there needs to be sustained attention to the role that schooling could play in contesting racism. This chapter underscores the need for alternative pedagogical strategies that extend far beyond curricular supplements, add-ons, and intercultural guidelines (Roman and Stanley 1997), something akin to a "radical redefinition of school knowledge from the heterogeneous perspectives and identities of racially disadvantaged groups—a process that goes beyond the language of 'inclusivity' and emphasizes relationality and multi-vocality as the central intellectual forces in the production of knowledge" (McCarthy 1993, 290).

This chapter, like others in this section, highlights the extent to which intercultural education and anti-racist educational interventions at the school level are shaped by inter/national political-economic forces that extend far beyond the classroom. Vertical case studies employ a range of techniques, such as discourse analysis and ethnographic investigations, to consider the ways in which shifting political-economic arrangements become "charged and enacted in the sticky materiality of practical encounters" (Tsing 2005, 3). Similar to Valdiviezo's study of interculturalism in Peru and Ghaffar-Kucher's work with Pakistani immigrant youth in the United States, my work considers the ways in which teachers and students appropriate policies and discourses, resulting in new cultural forms of engagement across difference.

Notes

1. As the name suggests, Travellers are a nomadic group for whom mobility is central to the maintenance of traveller culture. The precise origins of the Irish Traveller community are unclear, but Ní Shuinear (1994) notes that references to the appearance of Irish Travellers as a distinct grouping in academic works occurred in the 1890s. While the present black Irish population is predominantly of recent immigrant origin, a small, but significant, black (or "mixed race") presence has long existed in Irish society.

2. The source of this quotation is not provided to protect the identity of the school. The source is derived from a publication produced by a national representative body for statutory educational committees responsible for educational provision at various levels in Ireland. The term "international students" was typically applied to ethnically and/or linguistic minority students at BHC, irrespective of how long they had lived in Ireland or whether they had Irish citizenship.

Chapter 8

"Don't You Want Your Child to Be Better than You?"

Enacting Ideologies and Contesting Intercultural Policy in Peru

Laura Alicia Valdiviezo

The Study of a Policy of Contradiction

The implementation of Bilingual Intercultural Education (BIE) in Peru began in the mid-nineties, during a time of social, economic, and political crises exacerbated by a civil war of unprecedented violence that began in 1980 and ended in 2000. It emerged in response not only to this internal upheaval but also to the international Education for All (EFA) agenda that emphasized universal educational access and to a trend toward BIE fomented by bilateral aid agencies, such as Germany's *Gesellschaft für Technische Zusammenarbeit* (GTZ). Peru's BIE policy focused on providing education to one of the poorest and historically most neglected sectors in the country: indigenous populations. A new constitution in 1993 placed indigenous people, at least nominally, at the center of the government efforts to expand educational access and to democratize the country. The Peruvian government officially adopted BIE to manage cultural, linguistic, and ethnic diversity and indigenous revitalization, primarily among the country's six million Quechua speakers. Nonetheless, as I have argued elsewhere (Valdiviezo 2006), BIE policy was less a national educational response than it was an element in the country's foreign policy;

therefore, while the BIE policy directed attention to an underserved population, it lacked fundamental information about the local communities to be affected and putatively helped by BIE. In Peru, this lack of knowledge constitutes the policy's major conceptual contradiction, namely, interculturality without sufficient attention to culture or cultural politics (see also chapter 7). In this chapter, I examine the irony of a national BIE policy created to combat the marginalization of indigenous language, culture, and ethnicity (largely the result of actions by dominant groups in Peru) by developing a too-often uninformed linguistic and cultural educational policy directed solely at indigenous populations. The analysis reveals that while fundamental contradictions in BIE policy channel curriculum and instruction in ways that contribute to the continued exclusion of Quechua culture and language in schools, thereby perpetuating the segregation of indigenous Peruvians, BIE policy has been appropriated by Quechua teachers who use the spaces provided by the policy to creatively address the challenges of diversity. Policies enroll diverse actors, who are forced to reckon with the policy, but teachers and other policy actors appropriate policies in ways that are improvised and often unpredictable (see chapter 1).

This chapter draws on a year-long ethnographic study that I conducted in three rural schools in the southern Peruvian Andes. It illustrates a vertical comparison of the conceptualizations and contradictions of official BIE policy and local teachers' responses to national policy in indigenous Quechua communities. I devote the first section of the chapter to describing the context of my study and my research design. Particularly, I situate BIE policy within Peru's complex political and socioeconomic landscape, where historical, local, national, and international forces collide. In the second section, I move into the ethnography and discuss how BIE practitioners appropriate official policy that has aimed at the educational access of marginalized peoples. Specifically, I analyze teachers' views on the purpose and impact of BIE and their reactions to parental attitudes toward BIE. I conclude by arguing that teachers' practices and beliefs can constitute a ground-up proposal for policy and educational transformation, an epistemology from below that can challenge the grave inequities afflicting indigenous Peruvian schools and the broader society.

The Shaping of BIE Policy: From Colonial Legacy to EFA

Before the adoption of intercultural policy in Peru, political-economic rhetoric consistently defined the indigenous as those who could not speak, read, or write in Spanish and as antithetical to the vision of economic progress for the nation. Several studies have shown the common institutionalization of exclusionary practices against indigenous populations, their languages, and their cultures (Mannheim 1984; Cerrón-Palomino 1989; Freeland 1996; French 1999; De la Cadena 2000). In this study, I look at educational institutions as the entities that have undertaken the historic role of *civilizing* the indigenous into imagined, modern, monocultural Spanish-speaking Peruvians. Similar to the history of native language eradication in British and Portuguese colonies (Bhatt 2005), in countries like Peru, this civilizing mission was clearly conceived through educational and language policy excluding the indigenous and serving a larger political-economic project of the modernizing state (López 1997; Hornberger 2000; King 2001).

Peruvian independence from the Spanish colonial rule in 1821 did not translate into fundamental policy transformation in the educational sector. Because of the prevailing conception that only a unified, homogeneous, Spanish-speaking nation could achieve economic and social progress, the management of cultural and linguistic diversity meant the attempt to eliminate indigenous ways of life. Therefore, from the early nineteenth century to the 1980s, educational institutions, in particular the rural public school serving indigenous populations, aimed at the eradication of indigenous languages and cultures through strict Spanish-only instruction.

This policy of socially whitening the nation through education was interrupted by a short-lived revolutionary government reform undertaken by the military socialist dictatorship (1968–1975). During this period, the government attempted to mandate the teaching of Quechua in all schools while it made the indigenous language of Quechua, together with Spanish, an official language in Peru. Such efforts were not fully endogenous: Bilateral donor agencies such as the German development agency GTZ supported indigenous education initiatives at the time. The government's changes stretched far beyond education (and beyond what some foreign and national elites were willing to accept) to include an agrarian policy that redistributed privately owned land to workers and the nationalization of industries, such as oil, that were largely foreign-managed at the time.

These radical changes brought international sanctions and the downfall of the military regime; soon, counter-reforms led by the national elite restored economic policies to privatize industry and attracted foreign investment while slowly reinstating social and educational policies. The end of the counter-reforms in 1980 marked the beginning of a two-decade civil war where the terrorist group, Shining Path, and the military led what has been acknowledged as the bloodiest and costliest episode in Peruvian history (*Comisión de la Verdad* 2003). By 1990, severe political violence had submerged the country in an unprecedented period of human rights violations mounted against the poorest sectors of the population, the indigenous non-Spanish speakers.

In this historical context, BIE became an even more appealing policy for the fragile Peruvian state because it simultaneously suggested to activists its alignment with indigenous rights and signaled to the international donor community that the government merited funds for recovery and reconstruction. In the 1990s, indigenous activists were pressing the state to redress the egregious human rights violations and continued sociopolitical exclusion of their communities. At the same time, an international consensus emerged around universal educational access, or EFA, as one path toward poverty alleviation and national development. In Andean countries, in particular, GTZ—an agency with an established presence in Latin America since the early 1970s—led the institutionalization, support, and implementation of intercultural programs for education reform between 1992 and 2007.

In 1996, GTZ started the project PROEIBANDES, with headquarters located in Cochabamba, Bolivia, serving the countries of Argentina, Bolivia, Chile, Colombia, Ecuador, and Peru. The agency provided technical preparation and consulting to ministries of education and universities while offering training of selected leaders in intercultural bilingual education. PROEIBANDES offered the Peruvian government a way to address the demands of indigenous populations for education pertinent to their culture and language and to educate indigenous leaders for political participation in the larger society (*Deutsche Gesellschaft für Technische Zusammenarbeit* 2008).

As Hornberger (2000) and Freeland (1995, 1996) have argued, state intercultural policies are influenced by external pressures and prevailing global agendas; however, such policies are differentially appropriated by national governments depending upon the social, cultural, and political context of policy reception. Similar to other Andean countries that received funding and technical support from the GTZ, the Peruvian government officially adopted BIE to manage cultural, linguistic, and ethnic diversity and indigenous revitalization. Over the years, official efforts to voice alignment with the international intercultural proposal have contributed to the government's more democratic political image and have facilitated much-needed financial support for educational and social projects that otherwise the government alone likely would not have undertaken. Thus, a political-economic motivation impelled the Peruvian government in crisis in the 1990s to adopt BIE policy. This adoption signaled the political alignment of the country with democratic inter/national agendas necessarily tied to economic development. This, however, is not sufficient to understand BIE as educational phenomena. As a vertical comparative analysis of the policy reveals, the government's conceptualization of BIE represents only one dimension of the policy; another dimension is the appropriation of BIE by educational actors in a specific setting.

Understanding BIE Policy across Time and Space: Methodological Approaches

My approach to national BIE policy utilized a retrospective, critical historical view as well as a sociocultural analysis of the policy's present institutionalization. In this framework, there is an understanding that knowledge of BIE policy is severely limited without acknowledging the impact of colonialism in the institutionalization of policy and social practice in relation to indigenous people. Indeed, the effects of colonialism are still felt in Peru, as control over laws and institutions is largely in the hands of dominant Spanish-speaking groups, while linguistic, racial, and cultural minority groups, such as indigenous people, have remained economically, socially, and politically subordinated. Thus, colonialism in this study constitutes a critical historical lens to interpret forces shaping how BIE policy is locally instantiated in the present.

By looking historically and vertically, my understanding of national BIE policy is grounded in sociohistorical perspectives and enriched through a sociocultural lens focused on the processes taking place locally, in bilingual elementary classrooms where policy is appropriated. Thus, I pay attention to the ways in which classroom teachers—myself included as a former Peruvian teacher—make sense of and enact education in the context of BIE policy. The role of teachers as policy actors is critical because the policy's stated aim is the revitalization of historically marginalized languages and cultures, which has the potential to bring local practices to bear on policy and to contest worldviews grounded in colonial perspectives that traditionally excluded local indigenous knowledge (Appadurai 2000; Canagarajah 2005). As Koyama shows in chapter 1, policies enroll actors, yet the directions of actors' responses are multiple, contested, and improvised. Teachers form a critical link in the translation of policy into practice.

Research Methods

My multisited ethnographic research, conducted from 2003 through December 2004, led me to various sites, including the Ministry of Education (MOE) in Lima, teacher-training centers in and around Cuzco, and three schools in the southern Andes. The main settings for my field-work research were three Quechua elementary schools and various other local settings, including professional development centers where the BIE educators attended in-service training. The three schools have been part of the BIE program since 1996, when BIE began implementation as a government initiative in this region. Each school is located in a Quechua farming community surrounded by areas where tourism is a primary generator of revenue. While tourism thrives in this region, even promoting the temporary hiring of Quechua men and boys as carriers, most Quechua people engage in agriculture and herding for domestic consumption as their main activity. Like indigenous people across the Andean nations, Quechua speakers in these communities have limited access to basic services and schools, where they have shown high drop-out rates and grade retention in an education system that has traditionally suppressed the use of Quechua language and cultural practices (World Bank 2005).

At the MOE in Lima, I collected data over several visits where I interviewed ministry officials from the BIE division and other related offices. For the purposes of this chapter, I analyzed Peruvian education laws and publications from the ministry, examining the definitions and prescriptions of BIE to identify official discourses regarding interculturality, inter-cultural education, and views of language, ethnicity, and culture. I then conducted eight months of fieldwork in three schools and six mandatory BIE teacher-trainings in the southern Andes. Overall, I interviewed (in Spanish) seventeen BIE teachers (eight female and nine male teachers), five teacher trainers, nine MOE officials, and five academics connected to the BIE program. Sixteen out of the seventeen teachers were bilingual in Spanish and Quechua. More than half of the teachers self-identified as native Quechua speakers. More than 40 percent of the teachers had 7 to 9 years in BIE, while approximately 30 percent of teachers had 3 to 4 years and approximately 25 percent had 1 to 2 years in the program. I spent roughly 10 weeks in each school conducting observations. The school observations took place during lessons, one-on-one teacher-parent meetings, or in school-wide parent events.

The Appropriation of BIE Policy: Analysis of Findings

From a comparative perspective that looks at multiple levels of policy and practice, the goal of understanding a phenomenon such as BIE situates the present inquiry in a context broadly conceived across time and space. In the preceding sections, I identified global and national forces, including EFA agendas and an unparalleled internal crisis that converged in a political-economic project that influenced the government's adoption of BIE as national educational policy in Peru. This relationship between national intercultural policy and global political-economic relations is similar to the case of Ireland discussed in the chapter by Bryan. However, my focus is on the teachers' appropriation of policy. In this section, I examine, first, the discourses in the BIE policy, paying attention to the contradictory ideologies of indigenous exclusion and inclusion. Second, I analyze how bilingual teachers appropriate BIE policy locally, in the context of indigenous elementary schools, focusing on how teachers' beliefs and practices both enact contradictory ideologies and signal a space for policy innovation.

Contradictions within BIE National Policy

Peru's adoption of intercultural policy has marked a significant shift in official discourse around indigenous populations, from one of national cultural homogenization to a discourse that "assumes that cultural, ethnic and linguistic diversity constitutes the wealth of the country" (*Ley General de Educación*, article 20, 2003). This significant shift is also evident in the present Peruvian Constitution, enacted in 1993, which emphasizes the country's cultural and linguistic diversity and the official status of indigenous languages (article 48). The general education law now defines interculturality as a principle of Peruvian education, based on "the acknowledgment and appreciation of differences and the disposition to learning from the other as well as on the harmonic coexistence and exchange between the diverse cultures in the world" (*Ley General de Educación*, article 8, 2003). Arguably, BIE official discourse emphasizes ideal notions surrounding the recognition of diversity and harmonious group coexistence while remaining fundamentally ahistorical, apolitical, and disconnected from the local social realities that BIE is supposed to serve. In this framework, official BIE policy stays away from challenging the status quo and the social practices that have perpetuated acute social, economic, and political disparities between sectors in Peruvian society and instead celebrates harmony and coexistence.

In the Peruvian education law, interculturality constitutes a universal principle that "ensures the quality and equity of education" (*Ley General de Educación*, article 10, 2003). Within this framework of opportunities for all, article 20 of the General Education Law also states that BIE takes into account diverse worldviews and practices by stating that it "incorporates the histories of peoples, their knowledge and technologies, their value systems and socio-economic aspirations" (article 20a, 2003). BIE's assertion of the inclusiveness of diverse worldviews is contradicted because official policy mandates that student evaluations in the BIE program respond to the national Spanish-mainstream curriculum that tests proficiency in Spanish as well as a host of standard school-based literacy and numeracy competencies. This is a curriculum which, inspired by notions of education as preparation to compete in the job market, has ignored indigenous languages and cultural practices (*Ley General de Educación* 2003; *Ministerio de Educación* 2003). Thus, mandated evaluations dismiss skills and resources acquired though everyday life in indigenous communities; instead, the evaluations measure content aligned with inter/national educational standards attuned with a global economic agenda of market competitiveness (*Ley General de Educación* 2003; *Ministerio de Educación* 2008).

The General Education Law establishes BIE throughout the entire education system in an attempt to make it comprehensive and consistent with a broader educational rights agenda. For instance, the Law "promotes...respect for cultural diversity, intercultural dialogue, the realization of the rights of indigenous peoples as well as those of the national and international communities" (*Ley General de Educación*, article 20a, 2003). From this perspective, as expressed in articles 20b and 20e, BIE "guarantees learning in the mother tongue, the learning of Spanish as a second language, and the learning of foreign languages," and, at the same time, [BIE] "promotes the development and practice of indigenous languages" (2003). Yet this global, rights-based discourse of linguistic diversity is called into question in that indigenous students are expected to learn Spanish while nonindigenous students are not required to learn indigenous languages (Hornberger 2000). Clearly, contradictions collide in BIE law when an education promoting the realization of the rights of all continues to position underserved populations as responsible for learning the dominant language while students from the dominant groups are not required to learn a language that would permit "intercultural dialogue." In a sense, the law suggests that the indigenous are responsible for undoing their own marginalization by learning Spanish while absolving the rest of Peruvian society from making equal efforts to learn the languages that have been marginalized.

In addition, the General Education Law urges the substantive "participation of indigenous groups" in the decisions, management, and development of BIE policy and curriculum (*Ley General de Educación*, article 20d, 2003; *Ministerio de Educación* 2003). However, BIE implementation in rural schools did not take place with indigenous participation, as BIE was instituted without prior information, dialogue, or consent from the indigenous community. Contrary to what the written policy states, BIE was conceived without local participation (Valdiviezo 2009). The absence of the participation of indigenous representatives indicates that the BIE policy was prescribed from the top rather than produced in collaboration with indigenous groups as one would expect from government documents. This fact suggests that the government has appropriated inter/national discourse of educational access for the marginalized while resisting the spirit of it by not engaging in "dialogue" over a policy with profound implications for indigenous children and their communities. The incorporation of intercultural, rights-based discourses has improved the government's democratic reputation internationally, thus facilitating foreign economic influx to support the country's development agenda; however, as shown in the next section, homogenization and indigenous exclusion continue in society and in schools.

BIE Policy in Practice: Teachers' Perceptions and Attitudes in Indigenous Schools

This section examines policy as instantiated in practice by highlighting the role of BIE teachers as sociopolitical actors who appropriate BIE policy. In this light, BIE teachers reproduce but also contest ideological forces that have excluded indigenous language, culture, and ethnicity, and, overall, indigenous people in school and society. While I discuss how education policy contradictions may contribute to the ultimate conception of the indigenous as antithetical to development, my analysis moves beyond the depiction of education institutions as mere mirrors of social disparity and indigenous exclusion. I place special emphasis on the BIE teachers' role as central to understanding school itself as the locus of forces of policy contestation and innovation.

"Why Do We Have BIE Only in the Rural Schools?": Teachers' Critique of the Policy Purpose

At a national level, we are not intercultural, that is why there is segregation and we name-call others: *cholos, serranos*. That segregation reinforces stigmas and affects the identity and self-esteem of our students. I ask myself, why do we have BIE only in the rural schools? BIE should be implemented also in urban areas because it is there where discrimination comes from. BIE should be implemented at a national level. (Interview, July 1, 2004)

During interviews and conversations, BIE teachers questioned the inequities they perceived behind the policies defining BIE as a program solely for rural indigenous schools. Bilingual teachers tried to make sense of BIE policy and its aim to reverse inequities by explaining that segregation in Peru exists because there are no intercultural relations. Most bilingual teachers were quick to identify widespread practices of ethnic discrimination, such as the common use of labels like *"cholos"* and *"serranos,"* to denigrate people of indigenous origin or people who come from the highlands or a rural location.

Bilingual teachers related their views on discrimination, an issue invisible in apolitical BIE official discourse, to their understanding of the need for interculturality, a need that, as suggested by the BIE teacher quoted earlier, reaches beyond the rural school to the broader society. Within this framework of discussion, BIE teachers often questioned the rationale behind the policy's emphasis on rural implementation of BIE. In some cases, BIE teachers seemed suspicious and referred to the government's discriminatory hidden agenda (Fieldnotes, June 2004). This agenda, some argued, "keeps groups of power in business, while they want [BIE teachers] to follow it even if it does not support the appropriate development of the community" (Interview, October 26, 2004). In the critical view of bilingual teachers, as a project for solely indigenous sectors, BIE seemed to hide economic interests that paid lip-service to international human rights and EFA agendas to keep funding flowing into government agencies, while doing little to address the issues of the indigenous community or to fight the mainly urban discrimination against the indigenous.

"What Is This? Apus Don't Exist!": Dismissing Intercultural Teaching Innovation of BIE

> Supervisors really don't know what BIE is. I notice that when they come to supervise or when they monitor me. When I was teaching about the *Pachamama* [mother earth] to the kids, the regional director came to my class and said to us: "What is this? *Apus* [mountain spirits] don't exist!" And I said to myself, "He really doesn't know." (Interview, October 27, 2004)

Although there was no formal BIE training on how to address cultural issues in the classroom, teachers creatively found ways to incorporate indigenous cultural practices and beliefs in their lessons. Teachers' innovative intercultural practices in BIE, however, remained unsupported at the official level even when BIE policy prescribed the incorporation of indigenous culture in the school curricula. Aside from the official opposition to the teaching of indigenous cultural features, evident in the earlier quote, teachers lacked overall program support for actual intercultural teaching, which revealed the grave contradictions of institutional resistance to the teachers' efforts and voice in BIE. With respect to teacher training in BIE, there was no know-how regarding the ways teachers might include Quechua culture in lessons. In fact, bilingual teachers complained that, while the Quechua language was taught in the BIE program, Quechua culture and epistemologies were not. In addition, BIE teachers observed that, in many cases, the official Quechua materials that the MOE provided were inadequate. As an experienced BIE teacher noted: "Several materials [had been] prepared by Spanish-speaking teachers and then translated by another person" (Interview, November 18, 2004). Curricular material translated from Spanish into Quechua may have been cost-effective to provide, but they did not integrate indigenous worldviews.

Both the bilingual teachers who were native Quechua speakers and those who spoke Spanish as their native language were critical of the absence of training, resources, and support for them to study the cultural variations in different Quechua communities and to bring this knowledge into their teaching. One veteran BIE teacher said about the Quechua communities in the area: "The culture from [here] is different from the culture from the neighboring community. . . . Through interculturality we learn to value other cultures in the way that we value our own. We want others to know about our culture in the same way that we can learn from other cultures" (Interview, October 19, 2004). To learn and value others the way interculturality proposed, to "learn about the culture and *cosmovisión* [worldview] of the community," BIE teachers referred to the need to develop research skills, something they did not receive through BIE trainings but clearly thought they needed to be effective in the classroom (Interview, October 27, 2004). Despite serious limitations, BIE teachers creatively utilized Quechua cultural practices in their classrooms. From incorporating indigenous rituals to emphasizing Quechua values in the classroom, teachers used content pertinent to the indigenous culture and adapted topics in the mainstream curriculum. Nonetheless, opportunities such as these for ground-up transformation of policy and practice in BIE remained unnoticed, if not opposed, beyond the indigenous classrooms.

Whether the pedagogies that the bilingual teachers used and the ways in which they connected their own understanding of Quechua cultures were the most appropriate was neither adequately challenged nor supported. Teachers' assertive practices and opportunities for policy innovation were dismissed, as shown by the regional director's rejection of Quechua views of *Pachamama* and *Apus* as nonsense, even though my observations of this lesson suggest that it was highly valued by the students and the parents involved (Fieldnotes, October 2004). The dismissal at the national level of these efforts to use indigenous worldviews as teaching resources means that BIE teachers' innovative ideas about BIE rarely transcended their schools and remained unknown, and unappreciated, at official levels.

"They Answer to You with Pride": Quechua Instruction and Student Success

> If you speak to first grade children in Spanish, they will not answer to you. They will only look at you, but when you speak to them in their language, in Quechua, they answer you. If you ask them: *"¿Iman sutiky?"* [What is your name?]. They answer to you with pride *"¡Nuqaq sutikay Antoniucha!"* [My name is Antonito!]. (Interview, November 15, 2004)

Bilingual teachers often attributed the success of their students, including student motivation to learn and to participate, to the use and validation of Quechua in the classroom. Moreover, the majority of teachers argued that the success of their students in the BIE elementary schools transferred to their achievement in Spanish secondary schools. Veteran BIE teachers frequently recalled how colleagues from the high school nearby praised the performance of their former BIE students (Interview, November 15, 2004).

BIE teachers also believed that through BIE, indigenous students developed confidence to face common attitudes of contempt and discrimination outside their community. A bilingual teacher recalled a school field trip to the ancient citadel of Machu Picchu: "I heard people said, "these kids stink, they are *wayrurus!*" [red beans or pits used in amulets; a derogatory term], but the kids and I responded to them...in Quechua "these children can be president someday!" (Interview, July 7, 2004). The use of the indigenous language in BIE not only facilitated students' academic success but also empowered them to challenge the inequities they could experience beyond their indigenous school.

"How Can We Be Equal to Those Indios?": Views of Parental Opposition to BIE

> There were parents [in our school] who did not want Quechua to be taught in the school and this worsened when one day we brought the

children for a BIE school wide competition [. . .]. We came with the
parents and the students to participate, and then [the parents] saw all
the other children from [the schools from the highlands] who were all
dressed in their typical clothes. For the parents [in our school] this was
like a bucket of freezing water, so they questioned us, "How can we be
equal to those *ch'uspas* [a bag to store coca leaves, here used as a derogatory term], to those *indios*? How can our community be equal? We live
far away from their town!" And even later during a community assembly the parents complained, "They have made us equal to those ones!"
(Interview, October 11, 2004)

Local practices, including parental attitudes as described in this
interview, reproduce discourses of indigenous exclusion that are embedded in BIE policy. When explaining these attitudes toward BIE, BIE
teachers noted the discriminatory attitude of some Quechua parents
who dismissed other Quechua speakers. According to the BIE teacher,
the parents in her school resented BIE because it equated them to those
who, because of the choice of their clothes and the place where they
lived, were perceived as *indios*. Like the terms *cholo* and *serrano*, *indio*
is a common pejorative term encompassing at the same time the stigma
of lowest socioeconomic condition, ignorance, and "lack of culture." In
the construction of the antagonistic urban/rural relation, *indio* constitutes the epitome of rural, a moniker that some Quechua parents did
not want ascribed to them or their children. Without exception, all
members of the rural communities in the region participating in the
BIE program spoke Quechua as their native language; however, the BIE
teacher in the quote above recalled the indigenous parents' outrage after
realizing that they were part of BIE together with communities that, in
their view, were more rural, more indigenous, and therefore were essentially *indios*.

On many occasions, bilingual teachers mentioned that indigenous
parents demanded that their children be taught solely in Spanish—and
even in English—as they believed that only in that way their children
would have the advantages necessary to succeed. In several instances, BIE
teachers explained that parents approached them to request that their
children be taught in Spanish as opposed to Quechua. For example, a
young BIE teacher quoted a parent's request after the first day of school:
"*¡No profesora, no se lo enseñe en quechua, enséñeselo en castellano!*" [No,
teacher, don't teach him in Quechua, teach him in Spanish!] (Interview,
June 8, 2004). When BIE teachers explained the difficulties they faced
implementing BIE, they often highlighted parental attitudes denoting
discrimination against particular expressions of indigenousness, such as
Quechua language, culture, and ethnicity, and even class and geographic

location. In those ways, local practices reproduced discourses of indigenous exclusion that are embedded in BIE policy.

Although critical of external inequities affecting their indigenous students beyond BIE and the communities themselves, bilingual teachers seldom questioned possible disparities or discrimination within the program, including those present in their own actions. Clearly, the use of derogatory terms associating indigenousness to stigmas of lowest socioeconomic condition, ignorance, or "lack of culture" is part of social practice in the Quechua-speaking communities surrounding the BIE schools. Interestingly, while showing awareness of exclusionary ideologies in the opposition of parents to the teaching of Quechua in BIE, bilingual teachers also revealed their own assumptions about the limitation of Quechua speakers and the stigma of backwardness, especially during their interactions with parents. For instance, during a school-wide parent meeting that was conducted in Quechua, a second grade BIE teacher addressed a parent who disagreed with the teaching of Quechua in the school. The bilingual teacher, who had been speaking in the indigenous language, shifted to Spanish to respond to the parent's complaint: "In school, you learned only in Spanish, yet you are talking to me in Quechua. Don't you want your child to be better than you?" (Fieldnotes, November 2004). Similarly, another BIE recalled her response to a group of parents who wanted the monolingual Spanish program in their school:

> They said that the old program was better, that they, themselves, had learned all in Spanish. But I explained to the parents that if, as they said, the old education was better than the BIE program, then "why are you where you are now? You should have been better off. This community should have progressed and you also should have become professionals or something better, but you are simple peasants." That is what I explained to [the parents] and they understood and agreed. (Interview, October 11, 2004)

Bilingual teachers' beliefs and reactions toward parental attitudes regarding BIE revealed these teachers' own ideological contradictions. As part of the BIE program, bilingual teachers needed to advocate for BIE and the value of Quechua instruction in the indigenous school. However, during their verbal encounters with parents who opposed the use of the indigenous language in schools, BIE teachers resorted to denigrating Quechua parents, telling them that without bilingual education indigenous people could not be professionals or "something better." Moreover, parental opposition to BIE discouraged bilingual teachers from including indigenous parents' participation—a prescription of BIE official policy—in school decision making. This suggests that teachers considered indigenous parents fit to be educated but not fit to be partners in their children's education.

BIE teachers showed awareness of discriminatory attitudes regarding the indigenous speakers in Peruvian society as a whole, but they also perpetuated exclusionary discourses in their own practice. For instance, bilingual teachers were reflective and critical about the impact and purpose of BIE policy, and they advocated for the use of Quechua as a tool for school success and social empowerment. As critical practitioners, BIE teachers questioned inequalities behind BIE policy and their own limitations regarding their knowledge of indigenous culture. Bilingual teachers also proposed more equitable policy and practice innovations. These included the extension of BIE to urban settings and the development of BIE teachers' research skills for their actual learning of cultures and worldviews of the communities they served. Despite these efforts, though, bilingual teachers' perceptions and reactions to parental attitudes toward BIE showed their own constructions and exclusions of indigenousness. Under these conditions, their practice of BIE resembled the top-down civilizing mission of the Spanish monolingual school in that they believed BIE would make indigenous children prosper and develop, unlike their simple peasant parents, who were considered "failures" by some teachers. In these rural schools, the beliefs and reactions of BIE teachers often replicated the contradictions and exclusionary ideologies present in official BIE policy discourse.

Concluding Remarks

Inter/national forces constituted by international educational policies such as EFA and the bilateral funding agencies, combined with an unprecedented internal political crisis, influenced the adoption of official policies of linguistic, cultural, and ethnic revitalization of indigenous peoples in Peru. In this context, BIE teachers become actors in the implementation of a policy that feeds into a political-economic project, with a colonial legacy of power founded upon systematic racial, cultural, and linguistic exclusion. But, in ways that are contradictory as well as unpredictable, teachers appropriate BIE policy in an indigenous context with specific social interactions and cultural resources that they use to both reproduce the inequalities and exclusion of indigenous people in society, and initiate spaces for policy innovation that challenge the status quo.

A vertical analysis reveals context-specific peculiarities regarding the adoption of intercultural policies and practices, but there are also important comparisons to draw. Similar to the chapters by Bryan and Ghaffar-Kucher in this section, my chapter accentuates the interconnection of global economic and political agendas, national economic, political, and social policies, and school-based educational practices. In Ireland, as in Peru, the adoption of intercultural education responds to a larger political-economic project that has both rebuilt the democratic reputation of an otherwise fragile government and attracted much-needed foreign capital to the country. The United States, however, has felt less need to signal such inclusion through federal policies; instead, as in Ireland, the United States has invested in multicultural curricula that fail to consider the broader political-economic structures that have intensified racism and what Ghaffar-Kucher, in the next chapter, calls religification. Interestingly, while in Ireland the management of linguistic and cultural diversity emphasizes the potential for immigrant populations to fuel Irish economic development, in Peru, the rhetorical emphasis has been on the linguistic and cultural revitalization of indigenous populations, not on their inclusion as an economic, social, or political force.

By examining how policy circulates among international, national, and local levels, it becomes clear that the implementation of BIE holds promise and poses problems for the goal of indigenous revitalization. Understanding how BIE teachers appropriate policy can shed light on the grave inequities afflicting Peruvian schools and society, even in the context of initiatives that are only partially serving the interests of traditionally marginalized peoples. The teachers' appropriation of BIE policy has responded to parents' concerns, students' desires, teachers' own linguistic capacity and life experiences, as well as forces transcending the local indigenous community. Their innovations recommend the extension of BIE beyond rural indigenous schools to address social structures and practices that continue to marginalize indigenous populations.

The concept of appropriation shifts a focus away from the deficits of policy implementation and toward educational inquiry and policy that can bring practitioners' innovations to light as an epistemology from below.

Chapter 9

Citizenship and Belonging in an Age of Insecurity

Pakistani Immigrant Youth in New York City

Ameena Ghaffar-Kucher

Much like the Iran hostage crisis of the late 1970s and the Rushdie affair of the late 1980s, the September 11, 2001 terrorist attacks on U.S. soil are a flashpoint for a particular generation, once again bringing Muslims into the limelight and reifying the position of Muslims as the Other. In tandem with increasing xenophobia across the United States (Vlopp 2002; Yuval-Davis, Anthias, and Kofman 2005), the events of September 11 have resulted in a resurgence of patriotism within the United States (Abowitz and Harnish 2006). Exacerbated, perhaps, by the war in Iraq and media attention to al Qaeda, Muslim immigrant youth are increasingly constructed as "national outsiders" and "enemies to the nation" (Yuval-Davis, Anthias, and Kofman 2005; Abu el-Haj 2007, 30). While there has been some popular press regarding the effects of September 11 on Pakistani/Muslim communities in the United States, until very recently, the ramifications of this event and subsequent developments on students in public schools have not been adequately explored (Abu el-Haj 2002, 2007; Maira 2004; Sarroub 2005, Sirin and Fine 2008). Yet what happens within the four walls of a school is a vital part of the larger picture of changing social relations: Public spheres such as schools often reflect and shape the relationships and tensions that exist in society at large. By reflecting on the socialization experiences of Pakistani immigrant youth in public schools, this chapter examines the ways in which transnational geopolitics have created a hostile political climate in the United States for Muslim students, and the ways in which increasing feelings of insecurity translate into distrust toward individuals that are (or assumed to be) Muslim. This phenomenon especially affects the everyday lived experiences of Pakistani-American and Pakistani immigrant youth as Pakistanis form the largest Muslim immigrant group in the United States (Powell 2003).[1]

Drawing on a three year multisited ethnography, this study explores vertically the ways in which Pakistani-American youth are positioned at the national and local level as outsiders, and how they position themselves in relation to their own notions of citizenship and national belonging. I argue that the "religification" of urban, working-class Pakistani-American youth, that is, the ascription and co-option of a religious identity, trumps other forms of categorization, such as race and ethnicity (Lopez 2003; see chapter 7, this volume). Furthermore, I contend that religification significantly influences the youth's identities, notions of citizenship, and feelings of belonging. As I will illustrate in the following sections, Pakistani-American youth in American high schools are drawn to their Muslim identity, not solely for religious reasons but primarily for political ones. The ways in which these youth construct themselves and are constructed by others has significant ramifications in regards to their academic engagement and socialization.

I begin by providing some contextual background in terms of the prevailing geopolitical environment, followed by a brief description of the methodology and research sites. I then move on to a discussion on transnationalism and its effects on notions of citizenship. Drawing from ethnographic data collected in a neighborhood with a large working- and lower-class Pakistani immigrant population in New York City, I describe the religification process that the youth experienced in their school. In doing so, I show how global and local politics intersect and shape the construction of the citizen and noncitizen.

Geopolitics and Pakistani Youth:
The Making and Unmaking of Citizens

The terrorist attacks on the World Trade Center in September 2001 drastically changed the lives of many Pakistanis in the United States. Newspaper reports described the disappearances of thousands of Pakistani men from neighborhoods across the country, particularly in major cities such as Chicago (Bahl, Johnson, and Seim 2003) and New York City (Powell 2003). In January 2002, the Department of Homeland Security (formerly Immigration and Naturalization Service [INS]) enforced the National Entry-Exit Registration System (NSEERS) as part of the USA PATRIOT Act; accordingly, noncitizen males from 25 countries that were "designated as threats to national security" were required to formally register with the Bureau of Immigration and Customs Enforcement (Kim, Kourosh, Huckerby, Leins, and Narula 2007, 3). Pakistanis made up the largest immigrant group of the 25 nations (most of which have predominantly Arab or Muslim populates) required to register (Powell 2003).

In the years since the World Trade Center was destroyed, additional events across the globe have focused international attention on Muslims. These include the July 7, 2005 train bombings in London, the riots in Paris in November and December 2005, the Danish cartoon controversy in 2006, and the alleged London plot to blow up airlines in the summer of 2006. Consequently, Muslims living in the West have repeatedly been positioned as "an alienated, problematic minority" (Werbner 2004, 897) and as "most cultural Other, inimical to 'Western' values and traditions in an essential 'clash of civilizations'" (Abu el-Haj 2002, 309). This increased surveillance of Muslims greatly affects Pakistanis because of Pakistan's role in current global politics: On the one hand, Pakistan has been a key ally in the U.S.-led War on Terror; on the other hand, Pakistan has been implicated in terrorist attacks due to the proliferation (and support) of extremist *Madaris* (religious seminaries) within Pakistan, where many terrorists allegedly received their training. The recent (2008) terrorist attacks in Mumbai have compounded such suspicions. In the face of these events, and by virtue of their religion and nationality, Pakistani immigrants have come under increasing scrutiny in the United States (Sperry 2004; Buchen 2008) and around the world. Thus, working- and lower-middle class Pakistani immigrants and their children, in particular, face the dual difficulty of being Muslim in a part of the world where they are viewed with suspicion and having less access to social and cultural networks due to social and linguistic barriers (Stanton-Salazar 1997). How, then, do Pakistani-American youth in the United States, growing up in an era of increased suspicion and surveillance of individuals from the Middle East and South Asia, cope with rising levels of hostility in their schools and in the nation? What effects does this hostility have on youth's sense of identity and belonging? Finally, in what ways does this hostility affect the youth's experiences in schools and hopes for the future? These are questions explored in the following pages as a picture of the religification of Pakistani youth comes into focus.

Methodology

This research heeds Marcus's (1998) appeal for multisited research in places that are "simultaneously and complexly connected" (51) as well as Kathleen Hall's (2004) call for multisited ethnography as a way to understand the cultural politics involved as immigrants work to be recognized as citizens of a nation. Hall argues for greater attention to the ways in which "immigrant statuses are defined and debated, citizen rights and responsibilities invoked, structural inequalities challenged, and cultural identifications created" within cultural politics in the public sphere (110). Multisited ethnography enables researchers to "illuminate the more complex cultural processes of nation-formation and the contradictory and at times incommensurate forms of cultural politics within which immigrants are made and make themselves as citizens" (108). Drawing on this approach, vertical case studies pay heed to the "ethnography of global connections," remaining cognizant of the ways that the "global" comes to exist as instantiated in local interactions (Tsing 2005, 1). In this multisited, vertically constructed ethnographic study, I discuss how geopolitical shifts shape the cultural production of immigrant youth's identities and solidarities in one community in New York City.

In addition to visiting the local community during the spring of 2004, between February 2006 and January 2008, I collected data at a high school that not only had the second largest number of recently arrived Pakistani immigrant students in its borough but was also located in close proximity to a large community of Pakistani immigrants. I visited the school three times a week over a period of eight months. I interviewed 17 youth, 12 members of school staff, 4 parental figures, and 3 South Asian-American community leaders. I also conducted focus group discussions, primarily with students but also with a group of (non-Pakistani) leaders of a local community organization. I engaged more informally with 47 youth during the course of fieldwork, and several of these youth were part of the 6 focus groups I held at the school. I was also able to collect several samples of student work, which included stories about their experiences in schools. On a few occasions in a language class with a high percentage of Pakistani students, I was able to assign students a topic for writing assignments. A great deal of descriptive observation was done throughout the data collection period, primarily in the school but also in the homes of participants and in the ethnic community itself. I often visited community leaders in their offices, had lunch with them at local restaurants, and became, as much as possible, an observer of my surroundings. Through these multiple interactions, I developed an appreciation for the ways in which local and global politics were shaping the construction of "citizenship" and influencing notions of belonging.

Glocal Forces: The Religification of Identity

Ethnic diasporas exemplify transnationalism, which, broadly defined, refers to "multiple ties and interactions linking people or institutions across the borders of nation-states" (Vertovec 1999, 447). This is especially the case for "newer" ethnic communities, such as the post-1965 Pakistani community in the United States. The growth of instant communication through telephone, e-mail, mobile phones, videoconference, and faxes makes daily contact across national boundaries an "experienced reality" for transmigrants such that migration is no longer a unidirectional process (Werbner 2004, 896). However, a transnational lens does not posit a set of abstracted, dematerialized cultural flows; rather, it pays heed to the constant changes in people's lives and their reactions to these changes (Nonini and Ong 1997). Using such an approach, I endeavor to examine the "friction" created when universalizing concepts like nation and religion become "charged and enacted in the sticky materiality of practical encounters" (Tsing 2005, 3). In a similar vein, Werbner (2004) observes that diasporas are culturally and politically reflective, that is, they are constantly changing and negotiating their cultural and political identities; as such, they "cannot exist outside representation" (896). Questions of identity and representation have thus become important ones to consider. Stuart Hall (1996) suggests:

> Identities are about questions of using the resources of history, language and culture in the process of becoming rather than being: not "who we are" or "where we came from," so much as what we might become, how we have been represented and how that bears on how we might represent ourselves. Identities are therefore constituted within, not outside representation. (4)

In recognizing the importance of such questions of identity, this chapter argues that while race may have overpowered ethnicity in the ascription of identities for other minoritized[2] groups (see Lopez 2003), in the case of Pakistani youth, religion trumps both ethnicity and race. This ascription of an Islamic identity over other "affiliations, priorities, and pursuits that a Muslim person may have" should be taken seriously because of the way in which such an essentialist view of identity affects the social positioning of individuals in the nation (Sen 2006). Certainly, race and ethnicity complicate the ways in which Muslims are "imagined"; however, in contemporary social processes in the United States, nation and religion seem to outweigh these factors. Issues of nationalism, citizenship, and belonging are particularly salient to Pakistani immigrants given the "politics of immigrant incorporation and those of the 'global war on terror'" (Abu el-Haj 2007, 300). Questions of nationalism, however, have remained conspicuously absent in the sociological literature on the cultural processes involved in "becoming an American." Kathleen Hall (2004) contends,

> The nation—the boundaries of which imply the very terms of distinction between migrant and immigrant—is reified as an enduring context within which the immigrant experience takes place. This reification of the nation and of nationalism limits our ability to explain fully the cultural dynamics of immigrant incorporation. (109)

The problem with holding ethnicity or national identity static is that while it is telling of how people make sense of their lives, it provides little explanation as to how such "classificatory schemes are produced, circulate, and organize social practice" (Hall 2004, 112). In the case of Pakistani youth, the events of September 11 have significantly influenced questions of citizenship. As Yuval-Davis, Anthias, and Kofman (2005) argue, the mounting surveillance of individuals who are seen to be "different" after September 11 has "increased racialization of 'Others' in Western countries and affected constructions of national boundaries" (517). However, my research suggests that religion is used to define and categorize Pakistani youth (and their families) in the United States more than race and ethnicity. I refer to this as the "religification" of identity. Although religification resembles racism, in that it is a social construction, I contend that it is in fact different from racism, as elaborated in the next section.

Religification in Action

The religification of identity works bidirectionally. The Pakistani youth in this study were essentially "stripped" of their other—national and ethnic—identities, such as that of Pakistani, American, South Asian, and Desi,[3] and their religion became the primary prism through which they were seen by peers, teachers, and members of the community. In effect, religion segregates Pakistani youth, by marking their similarity to other South Asian and Middle-Eastern students (who are all presumed to be Muslim) and their significant difference from most peers. Further, the Pakistani youth increasingly identified themselves as Muslim because this move allowed them to transform a negative experience of being ostracized as different into a positive one of solidarity and group membership with other Muslims both in their immediate community and globally.

However, the Pakistani youth with whom I worked also increasingly identified as Muslims not so much because of heightened feelings of religiosity but for more personal, and occasionally political, gain. At times, the youth co-opted a religious identity to find a community where they were easily accepted, and at other times they took advantage of the heightened awareness by school personnel of Islamic practices to gain certain privileges. For example, a number of boys in the study frequently cut school on Friday afternoons, using the excuse of Friday prayer that was unofficially sanctioned by the school. However, some boys never made it to their prayers and simply went to the park to play. How then did the Pakistani youth make sense of the ways in which they were viewed by their peers, teachers, and fellow Pakistanis, and how did they themselves contribute and react to the hostile climate surrounding them?

For many Pakistani-American youth in this study, September 11 was a turning point in both their academic and social lives because of the ways in which they were verbally (and sometimes physically) attacked and ostracized by peers and occasionally even by teachers. Despite steps taken by schools and other groups to increase awareness and understanding about Islam and to reduce the stereotypes surrounding Muslims, the Muslim terrorist stereotype continues to flourish. How the increased focus on Muslims in public schools influences the religification of Pakistani immigrants requires greater attention. Abu el-Haj (2007) observes,

> Schools play an important role in the construction of the symbolic boundaries of the nation—in constructing who is and is not a member of the nation—and in the provision of resources with which immigrant youth learn to belong and navigate their new society. (288)

From the perspective of the youth in this study, before September 11, no one really knew or cared about who Pakistanis were or whether someone was a Pakistani. As a 16-year-old Pakistani-American high-school girl that I interviewed put it:

> After September 11th, I think everybody started knowing who Pakis really are…before that it wasn't really common. No one really cared. (Focus Group Discussion, May 15, 2007)

Similarly, a 14-year-old Pakistani-American boy explained that before September 11, people were ambivalent about someone being a Pakistani; after September, in contrast, being Pakistani meant being Muslim, which meant being different:

> They see Muslim people as different. Like before there was just like regular people, you're from Pakistan... they don't think of it [being Pakistani] as a good thing, or a bad thing. (Interview, April 24, 2007)

One result of this "being different" was that many youth who had once gotten along with students of all nationalities, races, and ethnicities reported changes in behaviors of friends as a consequence of September 11. For example, Bano, a high-school sophomore who had been away in Pakistan on September 11, returned to school in 2002 to find that people looked at her differently: "When I came back, it felt like people looked at me differently. Even my old friends would say, 'you're like one of them too'" (Interview, May 29, 2007). In regards to increasing scrutiny that Pakistanis in general experienced after September 11, one mother reasoned, "Perhaps it was in our minds," suggesting that they themselves imagined a change in people's behaviors (Interview, April 3, 2007). Whether changes in attitudes were real or imagined, the fact is that for the youth, their lives changed. Their sense of exclusion translated into behaviors that only served to compound their outsider status (see also Willis 1981).

The feelings of "us and them" continued well past the events of 2001. Many students talked about how September 11 was a turning point in the way they experienced schooling in the United States from that point forward. The following are some of the written responses that students in 2006 gave to the question, "Did race, gender, ethnicity, and socioeconomic status affect your experiences at school? How?":

> My school experience was pretty good before 9/11. However, in sixth grade, after 9/11, I have been looked [at] as a stereotype. Everybody used to call me a terrorist. All through Junior high school, people used to call me Ossama or Saddam. (Latif, 15-year-old boy)
>
> At one point after 9/11 it really did [change] because of my religion and where I come from. The kids in my class used to call me "terrorist" but that really didn't bother me because I didn't care for what they said. (Yasir, 14-year-old boy)
>
> There was no concern for it [race, ethnicity] until 9/11. Then kids would talk trash and they would get beaten up. (Tariq, 16-year-old boy)
>
> Around 9/11 students use to bother me and tease me because I was the only Muslim there (except my brother who was teased too). (Ishra, 17-year-old girl)

For many youth, the ostracizing and negativity around being a Muslim that the youth experienced after 9/11 strengthened their identification as Muslims while widening the gap between Muslims and non-Muslims, thereby making the possibility of being "American" less attainable. This was poignantly articulated by a group of Pakistani (Muslim) girls in a focus group discussion:

Soroiya: Everything has a stereotype and there's a stereotype for Americans.
AGK: What is that stereotype?
Soroiya: Like being White.
Iffat: Italian.
Soroiya: Christian.
Marina: Definitely not Muslim. Ever since September 11th, definitely not Muslim.
(Focus Group Discussion, May 15, 2007)

Marina's statement, "ever since September 11, definitely not Muslim," is particularly telling. She had just started the seventh grade in 2001, and before the events of September 11, she had never wanted to associate with Pakistanis or even Muslims. Now, however, she identified herself as a Muslim. This was not so much because she was a practicing Muslim but because she felt that she could no longer claim an American identity. Marina, who was born in the United States, continued, "I feel weird saying I'm American because American people, they don't like Muslims so much" (Focus Group Discussion, May 15, 2007). The dislike for Muslims that the youth experienced ranged from verbal spats to physical fights. Almost every Pakistani immigrant youth I spoke with said that at some point during their post-9/11 years in school, they had been referred to as a terrorist.

Playing into the Terrorist Stereotype

The terrorist label was one that came up frequently in discussions with the youth and seemed to be one of the reasons why they started ascribing a Muslim identity to themselves. Boys were more likely to have been called a terrorist, but girls also described such experiences. As Marina recounted: "At first I did not fight so much, but I got into so many fights. Because we couldn't sit there while people called us 'terrorist, terrorist.'" She also explained that the ostracism she experienced after September 11 drew her to fellow Pakistanis and to a common Muslim identity that had not appealed to her before:

Before we came to this school, we mostly didn't talk to Paki kids or Muslims, we didn't even like being Muslim back then. We didn't even want to be Pakistani, we didn't want to be talking to our own kind of people, we just didn't like them. (Focus Group Discussion, May 15, 2007)

While many youth reported that fewer people ascribed the terrorist label to them by the mid-2000s, they continued to bring up the term during the period of my research. For example, during a focus group session in 2007 with primarily non-Muslim and non-Pakistani students, one (non-Muslim) girl explained how she had avoided all Pakistani immigrants before September 11 but now affiliated with them:

I wouldn't even look at them but now, like, I became friends with them, and I noticed that it's not their fault, that, you know, they're not all bad. You know. It's just that one group that makes everybody look bad. Like I know Marina, Marina is not a terrorist. (May 22, 2007)

Most Pakistani immigrants and Muslim students, however, were not given these kinds of second chances and students did not make the effort to "get to know" them. Even in this particular situation, the final comment about knowing that "Marina is not a terrorist" indicates that the suspicion of Pakistani youth being terrorists continued for years after September 11. It is highly unlikely that *any* of the students at the high school were terrorists, yet the terrorist stereotype still hung in the air.

While at times youth criticized the terrorist stereotype, at other times they co-opted and used it to suggest power over non-Muslims and a shared identity with one's fellow Muslims. The following excerpt from an interview with Walid, a high-school senior who arrived in the United States at the age of two, illustrates this:

Walid: People in school were scared of me . . . I said something bad to them, "Yo, I'm going to blow up your house."
AGK: Why did you say that?
Walid: Just to scare them. So I became like an outcast, a rebel, and it felt good.
AGK: It felt good?
Walid: I felt different . . . They threatened me, Miss. I didn't just say "I'm going to blow up your house." Only if they picked on me, then—"Oh I'm going to fight you outside! Oh yeah? I'm going to blow up your house." (Interview, June 12, 2007)

Walid's story was just one of many where Pakistani-Muslim youth felt cornered, and one way they dealt with this was to take on the ascribed persona of a terrorist. Although this was a more common tactic among the young men in the study, some of the girls also took this route to dealing with the hostile environment at the school, as one young woman described:

Now we just joke about it. We'll be like, "get away from us, or we'll bomb you. . . ." You know how you make a joke out of stuff because it bothers you

but you don't want to show that it bothers you. You joke about it, I guess….
(Focus Group Discussion, May 15, 2007)

The ascription of the terrorist label and negative behaviors to Muslim students was not limited to their peers. It was also evident in teachers' interactions with Muslim students. Talah, a high-school sophomore, relayed an incident in a classroom where an Arab student was asking his teacher how she came to school. The teacher apparently replied, "The Verrazano Bridge. Why, do you want to blow it up?" (Fieldnotes, May 7, 2007). Regardless of the intention of such remarks, they inflict symbolic violence on youth (see chapter 7) and remain on the minds of students for years.

Teachers also made remarks in jest or without apparent malice that, nevertheless, contributed to the climate in the school whereby Pakistani and other Muslim students were seen as terrorists. Returning to the focus group when one girl said, "Marina is not a terrorist," their teacher, who was also invited to be part of this group, jumped in and laughingly replied, "Oh yes she is!" (Focus Group Discussion, May 22, 2007). Although this was meant to be humorous, and the teacher who said this is greatly loved and respected by the students, these sorts of seemingly benign statements are what psychologist Derald Wing Sue calls racial microaggressions (Sue, Bucceri, Lin, Nadal, and Torino 2007). Microaggression is a contemporary form of racism that is invisible, unintentional, and subtle in nature, and typically outside the level of conscious awareness, but it creates a hostile and invalidating climate nonetheless. As a result, it can be just as harmful, if not more harmful, than outright racism.

This form of racism affected the socialization and academic engagement of many Pakistani youth. For example, Walid had experienced many overt and covert moments of stereotyping. In part, as a reaction to this, he purposefully provoked teachers in class. He often challenged teachers and would say that certain things were "against his religion," such as listening to music; however, when I asked if he really held such views, he replied with a laugh:

I just say that to get them mad. Because I like it when people get angry… I do it to a lot of teachers. (Fieldnotes, June 12, 2007)

This is just one example where the youth "used" their religion but did not necessarily mean that they were religious. In this example, religion helped Walid cope with the racism that he experienced as a young Muslim man though not in the sense of increased religiosity; rather, religious identification served as a defense mechanism. Although this may seem like a reasonable strategy under the circumstances, it creates a vicious cycle: Peers and teachers ostracize Pakistani youth, who react and provoke their peers and teachers, thereby reaffirming the initial stereotypical beliefs. This cycle, which often ended in the youth being punished, heightened these students' sense of victimization even in cases where they were the ones in the wrong. For example, Iffat, a sophomore who had arrived in the United States a few years before September 11, talked to me in an interview about her most "racist" teacher, who had failed her in summer school because she missed a few classes. She was unwilling to accept that her poor attendance was the cause of the failing grade, preferring to ascribe responsibility to the teacher's views about Muslims (April 24, 2007).

Bad Boys, Oppressed Girls

Reacting to the behavior problems of some Pakistani students, teachers often reminisced about the "golden era" of past students, who were described as being more deferential and quiet. Many teachers who had worked with Pakistani students for a few years talked about how the students had "changed." In Valenzuela's (1999) ethnography about Mexican-American immigrant youth, she writes, "Contemporary students, in failing to conform to this misty, mythical image of their historical counterparts, seem deficient, so teachers find it hard to see them in an appreciative, culture-affirming way" (66). In the same way, teachers in this study saw contemporary Pakistani students as having more academic and behavioral problems, which the teachers attributed to their length of stay in the country and to their religiosity. On several occasions, teachers used the words "narrow-minded" or "inflexible" to describe Muslim students in general but especially boys, which they attributed to religious beliefs. For example, one social studies teacher explained:

> The biggest problem that I have, and I have had, I think over the last couple of years with the [Pakistani] kids is, sometimes when there are political discussions in class, um, right or wrong, I find that the, the Muslim kids in general are very—they're not flexible to really listen to the other side of the story. (Interview, May 18, 2007)

In another social studies class, a teacher had wanted to show the students a picture of Prophet Mohammad. In this situation, I had interjected and cautioned the teacher not to do so as it could result in a very angry reaction from the Muslim students because depicting Mohammad in visual form is considered blasphemous (Fieldnotes, December 5, 2006). The potential for such a reaction seemed to confirm the teacher's stereotype of Muslims as "problem" students, while the possibility that this picture could be truly insulting was not taken into consideration by this teacher.

Although teachers often viewed Pakistani boys as problem students, Pakistani girls were sometimes seen as oppressed victims of Islam to be pitied. In their interviews with me, teachers frequently brought up the restrictions placed on girls by their families and equated this with Islamic tradition even though "sexual policing" occurs in many groups (Zhou and Bankston 1998; Lopez 2003; Purkayastha 2005). Beyond immigrants, Anita Harris's (2004) book *Future Girl* suggests that the regulation of the sexuality of girls is alive and well in the United States. Nevertheless, many teachers at this school believed that Pakistani culture, which was considered to be synonymous with Islam, is responsible for this policing.

As a result of this belief, several teachers assumed that many Pakistani girls would not be allowed to go to college and that they would be "married off" or would only be allowed to study in a short vocational program. Such viewpoints affected the kinds of guidance the girls were provided at the school. For instance, one of the guidance counselors told me that girls tried to stall graduation so that they could avoid marriage (Interview, April 16, 2007). In addition, one of the assistant principals told me: "I've had several girls who have been married off, I guess for lack of a better term. It's a reality. It's a different culture. You know, it's a different culture (Interview, May 18, 2007).

Despite these views, not one of the twenty-seven focal girls in the study indicated that she would be facing such a fate, nor were any of them trying to stall graduation. It may have been an issue in the past, but it no longer appeared to be a concern for the girls even though the perception remained that early marriage was a part of Pakistani "culture." The view of culture as static and impervious to change affected the ways in which these youth were treated in school. However, as the next section illustrates, notions of culture among the Pakistani community were also somewhat static.

A Question of "Authenticity"

I was able to gain a deeper understanding of the choices and decisions that the youth made—about marriage, schooling, and other matters—by visiting their homes and the largest Pakistani community in the city. The youth reported feeling significant dissonance in their efforts to be an "authentic" Pakistani, or (more often) an "authentic" Muslim, while also being teenagers in an American high school. Because their parents were, for the majority, quite religious, the youth's religious identities were entwined with national ones: For most, being an "authentic Pakistani" required simultaneously being an "authentic Muslim." Despite their working-class backgrounds that made travel to Pakistan infrequent, if at all, the youth maintained strong ties to Pakistan. Further, given the configurations of culture and nation in the United States, wherein middle- and upper-class white Anglo Saxon Protestants get marked as the norm and others are at most hyphenated Americans (Alba and Nee 1997), these youth rarely entertained the idea that they could be American without giving up their "Pakistani-ness" or "Muslim-ness." For the youth in my study, being an "authentic" Pakistani meant resisting all things American, or what the youth and their families referred to as "Americanization." Though the definition of Americanization varied among youth and their families, the one common theme was the lack of religiosity, sexual promiscuity, drinking, and taking drugs. As a result, youth felt caught between two contrasting ideologies—the ideology of being Pakistani and the ideology of being American. For instance, Walid explained that it was extremely difficult for him to reconcile his Muslim identity with an American one:

> It's like on one side you're pushing religion and on the other side being American, but you can't have both. Being Muslim is hard. Our parents never told us that being Muslim is going to be easy. You have to hold on to that heritage and not let it go. (Fieldnotes, May 7, 2007)

Although being Muslim and being American was viewed as static and in binary opposition, in fact, the conflicting expectations that these youth faced were not conflicts between two distinctive "bounded" cultures; rather, they were tensions surrounding the different expectations for them as Pakistanis and as Americans (Hall 2004). For most of the youth, the expectations of "tradition" associated with "being Pakistani" were generally stronger than the expectations of "modernity" associated with "being American." Negotiating these different values and expectations was not easy for youth, and many of them attested to this struggle. As Marina explained:

> I didn't understand what side you're supposed to be on or anything. Like you know, on one hand you're Muslim and they're saying, "You're Muslim, go this way"; on the other hand you're American and you have to be like this. Like if you go to the American side, they're never going to think of you as American, but if you go to the Muslim side, you're not Muslim enough. (Focus Group Discussion, May 15, 2007)

Thus, being Muslim even fell outside of the more common race/ethnicity categorization that is characteristic of American society. Soroiya summed up this feeling of not knowing where she and her Pakistani friends belonged: "We're not Muslim enough and we're not American enough" (Focus Group Discussion, April 3, 2007).

Religion was thus a way to connect to a larger Muslim community without "losing" one's heritage and culture, which many youth deemed incompatible with American culture. This identification did not necessarily increase the religiosity of the youth, but it did increase the importance of adopting a Muslim identity. For example, Rashaad, an academically engaged 14-year-old who identified himself as "Muslim," explained that he did not consider himself religious, but that most of his friends were Muslim (though not necessarily Pakistani). For him, having Muslim friends, even if not religious himself, gave him a sense of belonging and camaraderie with other youth who also suffered in the wake of 9/11. Thus religion also became a way to "mediate the vicissitudes of class and race" in the United States (Maira 2002, 138).

Conclusion

This chapter demonstrates the ways in which Pakistani youth encounter and respond to racism in a post-9/11 world. The religification of identity has serious ramifications for both the academic and social arenas for Pakistani-American youth and the ways in which these youth are constructed as "non-citizens." Certainly, in the case of the Pakistani youth in this study, global politics affect the way these students are constructed as outsiders to the nation, but the youth themselves also contribute to the local and national milieu by the ways in which they construct their identities in a religious, and frequently oppositional, light. As observed by Purkayastha (2005) in her work with South Asian Americans:

> [the] ideological fragmentation of their identity into religious versus national identity (for example, Muslim versus American) simply did not reflect who they were.... Consequently, the South Asian Americans felt that they were forced to defend their religion actively, and this conscious effort makes religion more significant (though not in way described by the critics) in their everyday lives. (39)

This study also shows that religion is increasingly being used as a formal marker of difference by the mainstream in the United States and by ethnic-religious groups themselves. As a result, Pakistani-Muslim youth see themselves as outsiders in the United States because of the religification they encounter in their daily lives, but they imagine themselves to be part of a "Muslim community" that provides an alternative identity as an "insider" even though this feeling of belonging is in a community that is increasingly being ostracized by the mainstream. It softens their "outsider" identity by giving them a place where they feel they "belong," even within an inhospitable environment.

This "othering" along religious lines is neither a new social discourse nor simply a phenomenon in the United States: Zakharia's case study of language policy in Lebanon (see chapter 12) and Bryan's examination of multicultural policy in Ireland (see chapter 7) both illustrate how such religification manifests in a variety of discourses, both at the national and local levels. What this volume suggests is that these discourses must be studied and understood vertically. As both the Valdiviezo and Bryan chapters in this section demonstrate, the political economy of diversity influences the official and unofficial conduct of inter- and multicultural education in schools. Furthermore, similar to other chapters in this volume, I contend that in an era where security concerns and uncertainty dominate, the cultural production of identity, particularly in school sites, develops in unexpected ways that can best be understood through vertical case study analysis.

The interplay of global politics, national political and ethnic ideologies, and local interpretations of race and nation in the lives of Pakistani youth has implications for their education, employment, and socialization in the years to come, and for US society more broadly. Returning to interview these youth in several years, as planned, will provide further insights into the durability of religification in their lives.

Notes

1. The majority of participants in the study were American citizens; a handful had permanent-resident status, and a few were undocumented. However, in this chapter, I use the term "Pakistani-American" for all three categories.
2. The term "minoritized" comes from Teresa McCarty (2004, 93), who asserts that the term minority can be stigmatizing and is often numerically inaccurate: "Minoritized more accurately conveys the power relations by which certain groups are socially, economically, and politically marginalized. It also implies human agency and the power to effect positive change."
3. This is a South-Asian term to describe individuals of South Asian origin.

Part 4

Managing Conflict through
Inter/National Development Education

Chapter 10

The Relief-Development Transition

Sustainability and Educational Support in Post-Conflict Settings

Mary Mendenhall

In recent decades, *sustainability* has become a keyword in international development circles and an overarching objective for work across a wide array of social projects, including education. Yet, no widely accepted definition or operational guidelines exist that adequately describe its meaning. Furthermore, much of the discussion about sustainability in international development discourse presumes a certain degree of social, political, and economic stability in the country or context in question. What, then, does sustainability mean in so-called fragile states?[1] In particular, what does the sustainability of educational support look like in the transition from humanitarian relief in an unstable emergency context to a more stable situation of economic and educational development? How is the concept of sustainability reconceptualized when programs originally developed by international development institutions for an emergency context are revised, elements discontinued, and fledgling national governments assume responsibility for the provision of educational services from international organizations?

The goal of this chapter is to answer these questions through a vertical case study that illuminates the range of opportunities, challenges, and contradictions facing educational sustainability in post-conflict contexts. To do so, it examines the efforts of the Norwegian Refugee Council (NRC) to sustain the *Teacher Emergency Package* (TEP) in Angola, both during and after the conflict that affected the country for 27 years. NRC's work with the TEP provides fertile ground for examining sustainability due to the longevity of the program in Angola, which spanned the acute emergency and post-conflict phases. Its work also permits an examination of the friction in educational development projects like this one that are themselves sustained by global connections among international donor organizations and national ministries and policymakers. Returning to this metaphor as discussed by Bartlett and Vavrus in the introduction, the interactions among the various policy actors in this case led not only to debate about the future of the TEP, but they also resulted in "new arrangements of culture and power" within the education sector in Angola (Tsing 2005, 5). As discussed below, NRC endeavored to transfer the TEP to the Angolan government and other international organizations as the country moved into a post-conflict phase, with NRC planning to withdraw from the country once the situation stabilized. The interactions necessitated by this transfer illustrate the changing dynamics of power among international organizations and between them and their national counterparts during the critical relief-to-development transition. In Angola, this friction affected NRC's ability to transfer the TEP in its entirety to the Angolan government, thereby compromising its sustainability but also creating new opportunities to integrate certain human, material, and physical resources into the recovering educational system. The resultant effect of this friction, I contend, can best be described as policy *bricolage,* a strategy the Angolan Ministry of Education (MOE) and relevant international organizations needed to employ in order to adapt, implement, and sustain TEP. As Koyama and Max illustrate in chapters 1 and 2, respectively, this process of improvisation is an underacknowledged part of educational policy studies. The subsequent chapters by Shriberg and Zakharia support my argument that *bricolage* is especially common in fragile states, whose post-conflict education policies are often "wired together" from the remnants of programs developed before and during an emergency. Together, these notions of friction and *bricolage* help to make sense of the ways in which inter/national development decision making is profoundly local in the sense that it, too, is inflected by situated interests that are amplified in post-conflict contexts.

Angola: Civil Conflict and Reconstruction

In 1975, Angola gained independence from the Portuguese and ended 14 years of warfare by the Popular Liberation Movement of Angola (MPLA), the National Union for the Total Independence of Angola (UNITA), and the National Front for the Liberation of Angola (FNLA) (International Crisis Group 2003). Despite shared goals to overthrow the Portuguese, these various groups never coalesced (Bethke and Braunschweig 2003). On the heels of independence, internal conflict ensued, primarily between the socialist, government-backed MPLA, which was supported by the Soviet Union and Cuba and led by José Eduardo dos Santos (president since 1979), and the rebel-led UNITA, which was backed by the United States and South Africa and led by Jonas Savimbi. The conflation of ethnic tensions, regional disparities among different ethnic groups, cold war politics, and other countries' self-serving interests in Angola's oil and diamond resources exacerbated the MPLA-UNITA conflict (International Crisis Group 2003).

Angola's natural resources were grossly exploited during the armed conflict, and the revenues from diamonds and oil filled both the MPLA and UNITA coffers. A corrupt wartime economy, involving the national oil company known as SONANGOL, created a financial system through which money flowed but never appeared in the accounting records. According to Malaquias (2007), the oil revenues disappeared into a "'Bermuda triangle' comprising...SONANGOL, the president's office, and the central bank" (230). One study concluded that, during the last five years of the war (1997–2002), at least $4 billion simply disappeared (IMF 2003).

Although widespread fighting has not resumed since the end of the war in 2002, the corruption that was endemic to the 27-year civil conflict in Angola continues today. Transparency International's Corruption Perception Index (CPI) ranks Angola 147th out of 179 countries, with an overall score of 2.2 on a 0 (highly corrupt) to 10 (highly clean) scale. Despite being categorized as a fragile state due to its history of conflict and accompanying corruption, Angola's oil resources have attracted the interest of investors. China, in particular, has invested heavily and granted the country billions of dollars in loans for which credit was secured against future oil production (LaFraniere 2007). Despite these growing investments and the country's abundant natural resources, postwar conditions have not improved significantly for the average Angolan citizen.

As a fragile state, Angola faces significant challenges providing basic services, including education, to its populace. Such states are characterized as countries with a "lack of political commitment and/or weak capacity to develop and implement pro-poor policies," and they often face particular challenges in development projects (Rose and Greeley 2006, 1). This is because some international development organizations avoid fragile states due to their real or perceived instability while others may exert undue influence on the country's decision-making process, making it difficult to discern national from international priorities. Notably, there is growing interest by the international community in the role that education can play in mitigating state fragility and contributing to peace building, which is leading to greater research, advocacy, and policymaking on this topic (Conflict and Education Research Group 2008).

State of the Educational System in Angola

Adding to the complexity of the fragile Angolan context is the country's limited educational infrastructure. In fact, the country never had a universal education system. Before independence, education was provided almost exclusively to Portuguese immigrants (Bethke and Braunschweig 2003). Despite efforts by Roman Catholic and Protestant missionaries, educational opportunities for Africans were minimal due to the isolation of missionary work from centralized colonial activities (Samuels 1970).

After independence in 1975, the nation was consumed by the civil conflict with UNITA and was unable to focus attention on already limited educational opportunities. The MPLA made some effort to offer education in its controlled and often less affected areas, and UNITA attempted to create its own curriculum and offer some schooling. Nevertheless, the high percentages of children among the internally displaced populations and those recruited by both sides in the fighting greatly hampered children's education (Bethke and Braunschweig 2003). The MPLA and UNITA's overall neglect of education during the conflict highlighted their obsession with the financial benefits of controlling the country's natural resources (oil for the MPLA and diamonds for UNITA) rather than human development.

At the end of the war in 2002, fewer than half of Angola's children and youth had access to the country's education system (Watchlist 2002). The civil conflict's damage and destruction of approximately 4,000 schools also stymied efforts to rejuvenate the education sector (ReliefWeb 2004). Government spending on education in Angola has been one of the lowest in Sub-Saharan Africa, with only 2.6 percent of Gross Domestic Product and 6.4 percent of total government expenditures allocated to education (UNDP 2007). Many teachers were semi-literate, underprepared, and underpaid, and they relied predominantly on teaching methods that promoted rote memorization and a teacher-centered pedagogy.

The MOE in Angola launched a reform of the general education system in 2004 in an effort to respond to the country's major educational challenges. These included unequal access to education for learners; very limited educational infrastructure (e.g., classrooms and schools); insufficient quality and quantity of teaching personnel; insufficient supply of equipment and instructional materials; weak administrative and pedagogical management of teaching institutions; and the related devaluation of the teaching profession (Grilo 2006). UNICEF and other international organizations helped to shape the educational reform in an effort to closely align the objectives with the educational targets outlined in the Education for All and the Millennium Development Goals. The implementation of the reform was envisioned as a long-term, phased process that would reinforce the right to compulsory primary schooling, reduce drop out rates, and construct and rehabilitate schools. Before these reforms and during the civil conflict, several international organizations, including Save the Children and Christian Children's Fund, were working in the education sector in Angola to provide educational support to children whose schooling had been interrupted. One of the most active organizations, in terms of its scope of work and tenure in Angola, was the NRC.

Inter/national Influences in Angola: The NRC

The NRC, a humanitarian nongovernmental organization (NGO) that provides assistance to refugees and internally displaced persons (IDPs), worked in Angola from 1995 to 2007 in an effort to respond to the humanitarian and early reconstruction needs of the Angolan population. During its 12-year tenure, NRC's humanitarian activities consisted of camp and transit center operations for refugees and IDPs, including the provision and reconstruction of shelter, latrines, and water wells; food and supply distribution; training of health care workers; advocacy, information, counseling, and legal assistance for returnees; and emergency education (Norwegian Refugee Council 2002). The crux of NRC's education work in Angola during these 12 years was the TEP. It entailed two core components: a teacher training program for under- and unqualified teachers, and a bridging program that granted children who were behind in or who had never been to school the opportunity to re-enter or enter the formal education system. The TEP had been developed originally by UNESCO's Programme for Education in Emergencies and Reconstruction (UNESCO PEER) based in Nairobi. UNESCO PEER developed the TEP for use in Somalia in the early 1990s amid the country's civil conflict; it was later used with refugees in Djibouti, Ethiopia, Kenya, Rwanda, Tanzania, and Yemen. The original package, developed by UNESCO PEER, consisted of a kit of didactic materials and a methodology for teaching basic literacy and numeracy in the first language of the learners. UNESCO PEER's TEP, described as a "school-in-a-box," was intended as a rapid educational response program during the humanitarian phase of a crisis.

Studying Policy Actor Friction and the *Bricolage* of Educational Policymaking

To generate a broader understanding of the ways in which educational support may or may not be sustained through the transition from a period of crisis to long-term development, this chapter draws upon a larger qualitative study that had two overarching goals. The first was to create a frame of reference at the international level about the critical factors that affect sustainability within the field of education in emergencies and post-crisis reconstruction by interviewing educational practitioners working in the headquarters' offices of different types of international organizations active in this field. The second goal was to examine, by using these critical factors identified at the international level, a specific example in a conflict-affected country of one international organization's efforts to sustain an education program in the transition from relief to development. The application of a vertical case study methodology for this research study facilitated a more holistic understanding of sustainability through the examination of its complexities in and across the municipal, provincial, national, and international levels as well as the range of organizational actors active in each level. Data for this study were collected between March 2007 and February 2008, and it entailed interviews with 12 representatives from international organizations located in Europe and the United States and involved in global emergency education efforts. It also included interviews with 33 key stakeholders associated with the TEP program at the local and national levels in Angola and visits to 3 provinces (Luanda, Cuanza Sul, and Zaire) to discuss the program with municipal, provincial, and national education authorities as well as inter/national NGO staff. The data presented in this chapter stem primarily from the fieldwork conducted in Angola.

Implementing and Sustaining the TEP in Angola: Challenges, Opportunities, and Contradictions

In 1995, with support from an education adviser at NRC headquarters, UNICEF conducted a feasibility study to determine whether the TEP was an appropriate program for Angola. At the conclusion of this study, the two organizations decided that NRC, with financial assistance from the Norwegian government, would become the lead agency in adapting the TEP to the Angolan context; NRC had also begun other humanitarian activities in Angola at this time, as stated earlier. During the adaptation process, an NRC staff member collaborated with a trainer from UNESCO PEER, staff members from the Angolan MOE, curriculum specialists from a local teacher training institute, and education colleagues from the UNICEF office in Angola (Norwegian Refugee Council 1998). This team of individuals worked to preserve the child-centered and participatory methodologies that were central to the original UNESCO PEER program, but they modified other aspects of the program both early on and throughout NRC's involvement with the TEP in Angola. These fundamental changes included the following: the six-month nonformal education program was extended to align with the formal education sector's nine-month calendar; the one-off teacher training sessions were extended over time to a modular, in-service format spanning seven weeks to respond to the needs of under- and unqualified teachers; a supervisory component was added in an effort to complement the teacher training modules and to provide ongoing in-service support to teachers; the target age group for the program was changed in 2002 from 10–13 years old to 12–17 years old to facilitate younger children's direct access to the formal system and to accommodate the enormous backlog of adolescents and youth who had been out of school; and traditional Angolan stories and texts were collected to supplement UNESCO PEER materials in different subject areas. In addition to these modifications, the program also responded to changes brought about by the Ministry's own education reforms that sought to augment the curriculum with the inclusion of topics prioritized by the Ministry and its international partners, including human rights and HIV/AIDS. Over time, NRC was able to gradually transfer responsibility for compensating the TEP teachers and supervisors to the Angolan government, though, as Shriberg discusses in the following chapter, the compensation may not have been adequate to meet teachers' needs. The Angolan MOE's involvement from the beginning with the adaptation of the TEP—a critical exercise for generating ownership of the program in the early stages—and the later transfer of responsibility to the Ministry for compensating the teachers and supervisors, were both significant accomplishments for ensuring the longer-term sustainability of the program and its related human resources.

NRC and the Angolan MOE: Shifting Priorities

Given the magnitude of the TEP program, NRC understood the need to cultivate strong relationships with the MOE at the national, provincial, and municipal government levels, and it actively sought to build them throughout the course of NRC's tenure in Angola. A strong relationship with the Ministry was vital because the overarching goal was for the government to assume responsibility for the TEP as a way to sustain it. In a process of interactional improvisation, the relationships and modes of collaboration with the Ministry, as well as other international organizations working in Angola, manifested in different ways, at different levels, and at different times. The friction among these inter/national actors and organizations kept the TEP transition moving, though at times it facilitated and at other times it hampered the policy reform process.

The opportunities for NRC to develop supportive relationships and to instill a strong commitment to the TEP within the MOE proved to be most successful at the provincial and municipal levels. The exposure of the provincial and municipal education authorities to the TEP provided these individuals with multiple opportunities to observe and learn about the program's methodology, which led to requests by Provincial Departments of Education for assistance to train teachers and supervisors from the formal education system. One staff member from NRC headquarters stated that at one point: "The Ministry discovered that the children in TEP were in many cases performing better than the same-aged children who had been in school regularly... [and they] discovered that there must be a reason for this. The differences were the methodology and the attitude of the teachers, etc. Then NRC was asked to train all first and second grade teachers in two provinces" (Interview, April 15, 2007). As a result of this request, NRC staff felt, "The provincial education authorities ha[d] obviously understood the participatory methodology and the importance of follow-up and supervision, and they appreciate[d] the capacity that ha[d] been built in their areas" (Interview, April 15, 2007).

Due to the longevity of the TEP in Angola, in many cases education authorities were found at the provincial and municipal levels who began as teachers or supervisors and had participated in the TEP training, thereby bringing their understanding of and training in the program to their work within the Provincial and Municipal Departments of Education. Other participants interviewed at the provincial level had performed dual roles in the formal education system and the TEP, and they commented about the cross-fertilization that occurred. As one staff member from an international organization affiliated with the program pointed out, the result is that "it's very easy to see [the TEP] as a government program" (Interview, March 21, 2007).

The TEP training proved to be such a success that in some instances the provincial education authorities immediately recruited teachers upon completion of their TEP training. Although this was a boon for government education officials, it did hamper NRC's own goal of expanding the program. According to one NRC staff member, this "abrupt absorption of teachers trained in the TEP into the regular schools prevented [NRC from] reaching the goals foreseen in the project" (Interview, October 15, 2007). Nonetheless, the hiring of TEP-trained teachers was also a clear indication of its success. The widespread appreciation of the TEP methodology at the provincial and municipal levels, and the education authorities' growing interest in and requests for training, created opportunities for the TEP to seep into the formal education system.

Although there seemed to be strong and continuous recognition of the TEP's contributions to the Angolan education system at the subnational level, this support was less secure at the MOE. As the country transitioned from a state of conflict to a more stable, development-oriented context, participants noted a decline of support for the TEP at the national level. They identified two different but somewhat related reasons for this change: one, the Angolan government's shifting orientation from decentralized to centralized decision making; and two, the challenges for international organizations to maintain personnel, especially people with the requisite skills needed to carry out a particular type of work.

Regarding (de-)centralization, the TEP was initially endorsed by the national MOE, and the NRC (in collaboration with UNICEF and the Ministry) was able to expand the TEP over time to 12 of the 18 provinces in Angola. During the civil conflict, the provincial and municipal education authorities were able to make fairly autonomous decisions about the implementation of the TEP in their respective areas. However, after the war had ended and the national government grew stronger, a movement toward a more centralized education system appeared to be underway.

One important example of the government's shift toward a more centralized system entailed its development of a new literacy and accelerated learning program (ALP)—*Alfabetização e Aprendizagem Escolar*—in 2007 for overage students who had missed out on formal schooling (*República de Angola* 2005). The ALP consisted of six years of schooling compressed into three with the opportunity to transfer to secondary school upon successful completion of the program cycle. The program presented older students the opportunity to complete a full cycle of primary education. Although the program had been designed for adolescents and youth more than 15 years of age, students younger than 15 were allowed to participate in the first module of the program and then transfer into the appropriate grade within the primary school system. In the development process for the ALP, the Ministry sought out NRC's TEP expertise. In fact, a staff member from an NGO familiar with the TEP who attended one of the planning sessions in Luanda noted that all of the materials distributed by the Ministry to the participants during the meeting were merely a repackaging of the TEP materials that had been supplied during the developmental stages of the government's program.

Though they used TEP materials, experiences, and human resources, the Ministry did not take into consideration all of NRC's suggestions for ALP. The main source of contention between NRC staff and the Ministry had to do with the number of hours dedicated to classroom instruction. The government's ALP called for 12 hours of instruction per week compared to the 25–30 set forth in the TEP.[2] This caused much concern not only for NRC, but also for UNICEF, both of which acknowledged the contradiction of trying to launch an accelerated program that condenses six years of schooling into three with fewer hours of weekly instruction. NRC and UNICEF expressed their concerns to the Ministry and tried to advocate for a review of this issue, but the national Ministry responded by stating that the reduction in hours stemmed from the lack of available classroom space to accommodate the target number of adolescents and youth they hoped to reach with this program—approximately 100,000 students per year (Interview, April 2, 2007). Some international organization representatives felt that their objections made it very difficult for them to discuss the ALP with the Ministry, and some NGO staff members suggested that the Ministry's interest in reaching a large number of people before the upcoming elections trumped the need for quality education (Interview, February 13, 2008).

Although NRC, and other international organizations, supported the MOE's efforts to take the lead in designing and implementing new education policies and programs, there was also concern about the sustainability of the TEP, especially in terms of the teachers, supervisors, and trainers affiliated with the program. A representative from the national MOE who participated in this study stated that the TEP would figure centrally in the new ALP. She noted that the Ministry already recognized teachers and supervisors affiliated with the TEP as well as the materials that had been developed over the years. According to her, the TEP experience would be "completely absorbed both in terms of human resources and material resources" (Interview, April 2, 2007).

For this reason, the Ministry's plans to incorporate TEP resources into the ALP were considered positive actions by NRC because they showed evidence of sustainability, at least for the human resources developed through TEP. However, it was unclear how quickly the new program would be implemented because core components of the education reform that were initiated in 2002 (i.e., didactic and administrative materials) had yet to reach significant portions of the country. The slow pace of the reform did not suggest to NRC staff that the ALP would be implemented any more rapidly. Considering that more than 50 percent of Angola's population is younger than 30, the implications of delayed or lost educational or equivalent vocational training are enormous (Grilo 2006). Therefore, a number of provincial and municipal education authorities wanted to continue the TEP to meet the needs of youth in their regions. Yet the centralized decision making meant that these local officials had to abide by the Ministry's decision to focus solely on the new ALP program.

Interviewees believed that another reason for the Ministry's wavering support of the TEP had to do with international organizations' own internal personnel-related challenges. Staff turnover can be pervasive in international organizations working in conflict-affected contexts due to the demanding nature of the work. It proved to be a serious challenge for NRC and UNICEF, with implications for stable relationships with education authorities. As one participant put it:

> The problem is that there hasn't been continuity with the people working there [in the capital] with this program on the NRC side. They have been changing and all these educational coordinators have different ideas. Being head of office at NRC in Angola there are many other things that you have to pay attention to so maybe they aren't paying enough attention to this program. (Interview, November 7, 2007)

Although staff turnover inhibited interaction and communication with the national MOE, there was also the challenge posed by the non-Angolan staff who lacked skills in Portuguese and relevant cultural knowledge to interact effectively with education authorities. According to one participant who had interacted with the various NRC expatriate staff in Luanda over the years, NRC staff members' proficiency in Portuguese may not have been strong enough if they wanted to be able to advocate and negotiate effectively with education authorities, most of whom did not know English or Norwegian. A number of other Angolans also suggested to me that NRC staff needed a better understanding of the culture in the MOE—and in Angola more broadly—to establish a presence, build relationships, and navigate an inherently complex environment. Whereas international NGOs can be highly effective in delivering services and building capacity, particularly at the community level, the ability of expatriate staff to cultivate ongoing collaborative relationships with education authorities at the national level presents a challenge. Longer-term engagement at the national level would allow international NGOs to better anticipate and assess changing political landscapes that may have the potential to influence the implementation and sustainability of their education programs. It may also be helpful for international NGOs to assist local education authorities who are collaborating with successful education programs in their efforts to communicate with the national MOE, thereby generating greater support for the program's ongoing implementation. Yet organizations such as NRC do not only need to build strong relationships at the local and national levels; they also work within a network of international educational development organizations whose members may support or oppose their in-country efforts. Such was the case in the relationship among NRC and UNICEF in their work on the TEP, which is explored in the following section.

Collaboration and Conflict between NRC, UNICEF, and the Angolan MOE

Throughout the development of TEP, NRC attempted to establish partnerships with a variety of organizations working in Angola. While several organizations contributed to the TEP over the years, NRC's relationship with UNICEF was particularly important vis-à-vis the topic of sustainability for two primary reasons: one, UNICEF plays an extremely influential role with the Angolan MOE, a role that had the potential to create as well as limit opportunities for other international organizations; and two, the organization typically has a longer-term presence than other INGOs, particularly those with humanitarian mandates, because UNICEF works on long-term development projects as well as shorter-term humanitarian ones.

After the first two years of TEP's implementation in Angola, the Norwegian government—the main donor for the TEP—requested that UNICEF and NRC collaborate more formally. The two organizations signed a memorandum of understanding (MOU) in September 1998 in an effort to coordinate TEP-related tasks and responsibilities (Norwegian

Refugee Council 1998). The MOU explained that NRC would be responsible for all training involving the teachers, trainers, and supervisors while UNICEF would be responsible for the implementation of the program. These implementation responsibilities consisted of collaboration with provincial and municipal education authorities regarding selection of teachers and supervisors; the integration of TEP students and teachers into the formal education system; distribution of TEP kits and materials for school construction; and, at times, transportation for TEP supervisors (Johannessen 2000).

In 2000, a tripartite agreement was signed between NRC, UNICEF, and the MOE in order to strengthen further the collaboration in regard to the TEP; however, the language of the agreement was loosely constructed and cited as a weakness in an evaluation of the TEP carried out in 2000 by an external evaluator hired by NRC. According to the evaluator, the tripartite agreement "describe[d] a division of responsibilities between the parties, but lack[ed] details on the areas and modes of cooperation" or on the "the hierarchy between the parties" (Johannessen 2000, 5). Despite the fact that NRC and UNICEF had initiated the TEP, the overall goal for the government was gradually to assume responsibility for this program in an effort to ensure the program's sustainability. The details for how and when the government would assume these responsibilities were never clearly accounted for in the agreement, posing difficulties for the sustainability of the program.

Although UNICEF remained involved to varying degrees with the implementation of the TEP over the years, numerous obstacles surfaced that proved frustrating to NRC and others collaborating on the TEP. Whereas NGOs tend to act quickly and nimbly in the coordination and management of their activities, UN agencies often become bogged down by their own bureaucracies (Reddy 2002). In the case of the UNICEF office in Luanda, this bureaucracy created delays for the procurement and distribution of didactic materials needed for the kit and the allocation of funds to partners involved in the implementation of the program (e.g., financial resources provided for teachers' accommodations and meals during TEP training sessions) (Interview, April 15, 2007).

Apart from the administrative delays that accompanied the collaboration between NRC and UNICEF, NRC staff were concerned about the ways in which UNICEF did or did not keep NRC informed about key information. In the early years, TEP coordinators were dissuaded from attending meetings that UNICEF scheduled with the MOE to discuss TEP. They were also not provided with updated information about the outcomes of UNICEF decisions that might have affected the implementation of the TEP.

There was a sense among the NRC staff that these challenges not only led to avoidable delays but that they also interfered with the provision of educational support. One NRC staff member felt strongly that the various administrative delays over the years had often "lower[ed] the intervention capacity of NRC and [created] some obstacles in the realization of capacity building courses for the teachers" (Interview, October 15, 2007). These challenges of collaborating with UNICEF prompted NRC to assume more responsibilities over the years, especially given the enormity of its invest-ment in this program. In contrast, UNICEF's priorities slowly shifted away from the TEP, so less attention was given to it. Considering UNICEF's influential relationship with the national MOE, its waning interest in the TEP may have negatively affected the relationship between NRC and the MOE.

The limited technical knowledge about the TEP among UNICEF staff due to extensive turnover also affected its ability to support the TEP adequately and collaborate with NRC effectively. Despite the presence of some highly qualified UNICEF personnel who were considered visionar-ies by some NRC staff and affiliated consultants, staff turnover at the UNICEF office in Angola and the inability of those who followed to engage at a deeper, more technical level with the TEP proved to be partic-ularly problematic. There also was agreement among those whom I inter-viewed that UNICEF may have lacked the vision as well as the interest for this type of project. The following statement spoke to the feelings among NRC staff that their efforts to engage UNICEF in the TEP were often fruitless: "You have some people from UNICEF who politely listen to you and then say, 'yes, we'll see,' and you knew not so much would happen. That was the kind of feeling that we got...you knew that they were busy doing [other things] and they didn't have the [human] resources to carry it through" (Interview, November 19, 2007).

Although the bureaucracy of UNICEF and the problems that accom-pany it can be found in other countries, the NRC staff with experiences working elsewhere in collaboration with UNICEF noted that the relation-ship with this UN agency in Angola was particularly problematic. The reasons primarily stemmed from the frequent changes in leadership in senior education management positions within the organization. This was a particular problem for NRC as it planned to withdraw from Angola: It needed UNICEF to be well informed about the TEP and the ways in which it had and might continue to inform the new ALP, particularly with regard to leveraging the vast human, material, and physical resources that NRC had created over the years. The lack of institutional memory about the TEP and NRC at the UNICEF office in Angola proved to be a serious limitation for sustaining the program.

Sustaining the TEP in Angola

In the case of Angola, the Ministry's decision to develop the ALP changed the country's educational landscape. It became less about NRC transferring an intact program to the government, as stated in the original goals for TEP, and more about transferring certain elements—the human and material resources—into a "new" national system, a *bricolage* of program and policy pieces that looked promising to the Ministry. Although an NRC staff member commented that the TEP is "only sustainable if the government wants it," she also thought in hindsight that perhaps a few additional steps could have been taken to ensure the sustainability of the program within the Ministry's education reform policies (Interview, April 15, 2007). NRC made several efforts of varying success to establish and fortify collaborative relationships with other international organizations such as UNICEF throughout TEP's implementation, but the interaction among them—the friction—inhibited new partnerships from being formed that could have contributed to the sustainability of TEP.

While the TEP enjoyed broad-based support within the Provincial and Municipal Departments of Education, this was not necessarily the case with the MOE as it grew stronger and sought to put its own imprimatur on existing programs. Moreover, the challenges of working with organizations that could have influenced the Ministry over the long term, such as UNICEF, highlight the difficulties in sustaining programs developed in the context of a humanitarian crisis as a country embarks on a more stable, development-oriented trajectory. Ultimately, the sustainability of the TEP depended on a complex series of improvisations and interactions between actors differentially situated in sites ranging from local Angolan schools through municipal and district educational offices to Luanda, Norway, New York, and beyond.

Concluding Thoughts on Managing Sustainability in Post-Conflict Contexts

The findings presented in this chapter capture several of the opportunities, challenges, and contradictions common to efforts to sustain inter/national development programs, especially in fragile states and post-conflict situations. In this case, the NRC's attempts to sustain the TEP collided not only with the Angolan government's efforts to establish its legitimacy and exert its independence during the post-conflict transition to development but also with the divergent goals of other international development organizations. Similar to Taylor and Wilkinson's chapters, decentralized governmental structures created opportunities for wider participation by local education authorities across Angola's provinces, in this case during a civil war. However, once the country stabilized and the national MOE began shifting toward a more centralized structure, the local education authorities' opportunities to continue their engagement with the TEP began to dwindle. As a result of these shifting governance structures, new interactional patterns emerged among international development organizations and with national ministries. This friction kept the transfer of TEP moving along, but the road along which it traveled was interrupted by the ALP. The resulting *bricolage* joined certain elements related to the human, material, and physical resources generated during the life of the TEP with policy reforms sought by the post-conflict MOE; the transfer of the program in its entirety to the Angolan government did not occur as NRC had intended.

Nevertheless, the recognition by education authorities of the positive contributions that NRC made to building the capacity of Angolan teachers, supervisors, and trainers augured well for these individuals' integration into the governmental system. This integration entailed the inclusion of the majority of TEP teachers on the governmental payroll during TEP's implementation as well as the ongoing incorporation of TEP teachers, supervisors, and trainers into the formal education system. It also included the utilization of TEP material resources on curriculum, pedagogy, and training in the development of the ALP. Finally, the exposure to TEP that education authorities at the municipal and provincial levels received enhanced their knowledge and understanding of the program and may very well influence their work in the future. The ways in which the organization implemented the TEP in a complementary manner in Angola and the subsequent integration and appropriation of the core components of TEP into the MOE's new ALP illustrates one example of what sustainability might look like in a country transitioning from protracted conflict to long-term development.

The vertical case study approach used for this study facilitated a closer examination of the various and evolving relationships among ministries of education, international NGOs, and UN agencies. It shed light on the importance of looking at these inter/national interactions as local encounters, profoundly shaped by the friction resulting from shifting political priorities and policy alliances, as further illustrated in the remaining chapters in this section. This chapter also demonstrated that policy influence does not move unidirectionally from the global to the national and local. Countries do resist as well as reshape policies and practices originating from international organizations, even in post-conflict contexts when states are often the most fragile, as chapter 12 by Zakharia illustrates. Moreover, the result is often a *bricolage* rather than a wholesale adoption of an intact program, which may better serve the long-term interests of the country. At the same time, initiating new policies and carrying them out may require ongoing assistance from the international community, a point that Shriberg's chapter to follow on Liberia makes vividly clear. In Angola and other fragile states, it is critical to explore effective ways to strike a balance between the assistance that international organizations can provide and the government's need to lead its country's educational reforms. Perhaps striking this balance in the post-conflict period rather than sustaining programs developed amid crisis should be the primary goal of international organizations.

Notes

1. The term "fragile state" is relatively new and generally refers to countries susceptible to conflict because of recent or anticipated domestic or international strife, and with very weak institutions to withstand or prevent such conflict (Crisis States Workshop 2006). The determination of state fragility is generally tied to a country's ranking in the World Bank's Country Policy and Institutional Assessments (CPIA); countries included in these assessments may or may not be affected by violent conflict but are seen as vulnerable to such conflicts. The following section provides further elaboration of the term in the Angolan context.
2. The number of hours of instruction for general education courses within the formal education sector total approximately 20 hours per week.

Chapter 11

Perpetuated Suffering
Social Injustice in Liberian Teachers' Lives
Janet Shriberg

This chapter asks how Liberian teachers in an early post-war context sought to maintain their own well-being despite the fragility of the Liberian state education system.[1] In accordance with the vertical case study approach, this research analyzes connections across international, national, and local levels that form during a humanitarian crisis and continue into the period of post-war reconstruction. Similar to Mendenhall and Zakharia in this section, I am interested in the friction within and among these groups of actors that arises from their "heterogeneous and unequal encounters" (Tsing 2005, 1). In particular, I examine how connections between inter/national governmental and nongovernmental institutions lead to the formation of policy that may inadvertently perpetuate suffering among war-affected teachers through very limited compensation policies. When these global policies about teacher compensation become "localized," first in national ministries of education and then in the lives of teachers, they create new forms of suffering for Liberian teachers that reinforce historical inequalities and marginalize teachers as professionals and as civil servants. The chapter begins by examining the policies of international institutions involved in the field of education in emergencies and shows how the connections among the Liberian Ministry of Education (MOE) have directly shaped post-war teacher compensation efforts. It then moves to the local level to show how extremely low salaries, the lack of housing, and the dearth of material and training support affect teachers' perceptions that they are being treated in a just manner by their government. In contrasting these local narratives with current inter/national policies of teacher compensation in fragile states, I argue that post-conflict education reconstruction programs may perpetuate forms of suffering and conflict if they do not respond to the particular needs of teachers more effectively.

Post-Conflict Education Reconstruction and Teacher Well-Being

The past two decades have seen increased international recognition of education as a universal right often denied to young people living in situations affected by conflict or other disasters. Such concern prompted the growth of programs in the field now established as Education in Situations of Emergency, Crisis, and Reconstruction, meaning education in areas affected by violence, human-made disaster, war, or environmental calamities. The formal development of the objectives and programs for education in conflict-affected regions grew out of the second Education for All (EFA) conference held in Dakar in 2000. At this meeting, 180 countries announced their support for the conference's *Framework for Action,* which ensured access to education for all children by 2015, particularly for those living in difficult situations such as contexts of displacement and/or war. Its proceedings led to the formation of the Inter-Agency Network for Education in Emergencies (INEE), an organization that brings together humanitarian organizations, ministries of education, universities, and United Nations (UN) agencies to address the education needs of children made vulnerable by violence and other disasters.

The INEE and its partners have enumerated how education can assist communities affected by disaster and/or human conflict. Education is said to promote community participation, health education, peace building, and opportunities for developmental and cognitive growth for children (Sinclair 2001; Tomlinson and Benefield 2005). In regions affected by armed conflict, schools can provide safe environments where structure, stimulation, and opportunities for healthy socialization with peers and adults can mitigate the trauma of war (Winthrop and Kirk 2005).

Central to the efforts of international and national education reconstruction programs are local teachers. Many teachers have been trained to communicate critical messages to children that can protect them from recruitment into fighting forces, sexual or economic exploitation, and exposure to increased risk of contracting HIV/AIDS and other diseases. More generally, teachers working in contexts of extreme hardship can still create a classroom climate that helps to support children's healing after traumatic events. However, despite their essential role in the process of education reconstruction, there is limited literature that describes the concerns and well-being of teachers working in conflict-affected regions. The literature that is available points to teachers' vulnerabilities, both physical and emotional, in situations of conflict and/or disaster (Sinclair 2001; Davies 2004; Masinda and Muhesi 2004). This literature reveals a lack of attention to teachers' material needs, low status, and poor pay (see, e.g., Asimeng-Boahene 2003; Bennell 2004). In what follows, I discuss the findings of my research with Liberian teachers working in the early period of reconstruction, specifically addressing how inter/national policies related to teacher pay and support may in fact perpetuate suffering and hinder efforts to foster peaceful conflict resolution in the country.

Methodology

In February 2006, during Liberia's period of early reconstruction, only weeks after President Ellen Johnson Sirleaf and her government were established, I initiated the data collection process for this research. I gathered qualitative and quantitative data over a seven-month period of fieldwork in three geographically diverse areas in Liberia: Lofa, Nimba, and Montserrado counties. Data collection occurred simultaneously in each region and included 66 interviews with teachers and policymakers, a survey of 106 teachers, 2 focus groups with rural and female teachers,[2] and participant observation in primary, secondary, and university classes, teacher trainings, parent-teacher association forums, and government-led interagency education meetings. During interviews, I employed a card-sort activity that enabled me to develop an ordered ranking of the issues that teachers perceived as affecting their well-being. Data were analyzed using both quantitative and qualitative approaches in a mixed-methods design. Descriptive statistics (i.e., frequencies and means) were calculated for the closed questions, and the responses were stratified by the demographic variables of interest (i.e., sex, age, geographic location, and type of school). Responses to the open-ended questions in the interview and surveys were analyzed by conducting a content analysis of all of the responses to identify emergent categories (i.e., the most important themes, such as "unmet needs"). All individual responses were coded into these categories based on set criteria. The frequencies of responses in each category (and, where appropriate, means) were then calculated. Finally, descriptive statistics were calculated to determine group characteristics, such as the teacher-participants' responses to survey questions about their teaching experiences summarized as groups by gender, age, geographic residence, and years of teacher training.

Fragile States, Inter/national Policies, and Rebuilding Liberia's Education System

Despite the global consensus that all children are guaranteed the right to attend primary school, education in fragile states remains one of the most difficult challenges of international development. According to Save the Children (2006), it was estimated that some 115 million children worldwide had been denied their right to primary education. At least 43 million of these children had been deprived of schooling because they lived in states affected by armed conflict. The combination of poverty, unstable and/or weak governance, and often-violent conflict such as civil war characterize state fragility in many low-income countries (Chauvet and Collier 2007). In this context, education is seen as playing an especially influential role in the security agenda because of its unique potential for peace building and "renewing the social contract between a government and its citizens" (Rose and Greeley 2006, 2). However, international donors may be wary of granting fragile states development assistance in areas such as education. Often, donors reason that dysfunctional policies and governance, which contributed to and continue to exacerbate a state's fragility, compromise the state's abilities to manage external investment properly. This situation is paradoxical: The very conditions that determine a state's fragility may impede its ability to receive donor support for development and/or post-conflict recovery. For example, there may be no agreement among or within donor agencies about providing money to cover local teachers' salaries, and the governments in these fragile states may not yet have sufficient funds or effective mechanisms for doing so themselves.

Although Liberia's war ended in 2003, its fragility continues. Fourteen years of civil war (1989–2003) and related conflicts with its West African neighbors resulted in the death of an estimated 270,000 people and the displacement of at least 800,000 of the country's small population of approximately 3.5 million (United Nations Development Program [UNDP] 2007). The psychological and social toll of the war was devastating and continues to affect children of school age and their teachers. Tens of thousands of children were abducted, tortured, and drugged to force their participation in the war as soldiers or sexual slaves for the fighting forces. Rape and gender-based violence against girls and women were routine weapons of war (Sirleaf 2006; Medeiros 2007).

The years of violence also caused the countrywide destruction of schools, hospitals, and roads, as well as damage to the government infrastructure in the capital city, Monrovia. The problem of widespread poverty that existed before the war was exacerbated by the complete collapse of the country's economic infrastructure during the war. Therefore, it is difficult, if not impossible, for the new government to simultaneously rebuild the educational infrastructure, create psychological support programs for teachers and students, and pay teachers a living wage.

The successful democratic election and the January 2006 inauguration of President Sirleaf were major steps toward Liberia's peaceful recovery from war. Immediately following her instatement, the president and her government asserted that central to their priorities in Liberia's early reconstruction were efforts to build capacity at all levels, including greater equity in the distribution of public services. As part of their plans, she and the relevant government ministries committed to supporting an education system offering broader gender and geographic access to people across Liberia.[3] This move may have come in direct response to the unrest that led up to the Liberian civil war, especially in poor, rural communities that have long been marginalized in an education system favoring the urban elite (Buor 2001; Levitt 2005). The stated mission of the Sirleaf administration to improve education was seen as a key element of peaceful development by international stakeholders. However, carrying out Sirleaf's mandate required the support and commitment of international donors to the newly developing Liberian MOE. Specifically, financial and human capacity was needed to restart an educational system that had entirely collapsed. An early post-war report by the UNDP in Liberia (2006) summarized the dire Liberian education conditions, stating that "over 75 percent of the educational infrastructure was either destroyed or damaged and some turned into military warehouses and war rooms" (45). The Ministry was operating in a dilapidated building that lacked electricity, computers, and often basic furniture and office supplies, leaving Ministry staff few resources with which to carry out their work. Further, the poor road conditions, lack of vehicles, and small budget resulted in limited travel opportunities for Ministry staff to oversee the countrywide program. The very limited number of Liberians with the skills to fill urgent positions in the Ministry was a further hindrance; years of interrupted education, displacement, and war had created a serious brain drain from the country, leaving few trained professionals able to work in higher level administrative. As an example, the Liberia Local Government Capacity Assessment found that "local governments can hardly perform 20 percent of their expected roles and functions" (UNDP 2006, 48). There was also a dearth of trained teachers in Liberia's early reconstruction period. The UN Consolidated Appeal for 2006 reported that only 20 percent of teachers in Liberian public schools were qualified to teach (Women's Commission for Refugee Women and Children 2006). Teachers had not been paid for years, reducing the incentive to return to teaching in the post-war period. The war severely weakened an already impoverished economy, thus diminishing the abilities of the new government to adequately distribute national funds to education.

In Liberia's period of early reconstruction, donor support for education was designated specifically for programs selected by UN agencies and certain international nongovernmental organizations (INGOs). Programs such as the Accelerated Learning Program for overage students (see Mendenhall, chapter 10) and vocational training programs for young adults received substantial external help from international donors. At the same time, the UN continued to finance and assist in the implementation of primary school programs, which included school feeding initiatives for students. In addition, international organizations were able to offer financial support for school construction, specialized teacher trainings, and psychosocial support to students made vulnerable by the war. However, financial assistance designated for teacher salaries was not initially included. Reluctance by external donors to support teacher salaries may be explained, in part, because teacher salaries were felt to be "the responsibility of the current government of Liberia, and [their assumption of this cost] is critical for stability and reduction of corruption in the education system" (Women's Commission for Refugee Women and Children 2006, 19). Therefore, the bulk of teacher salaries for government school teachers was supposed to be paid out of the national budget. However, the budget itself is extremely small and the portion allocated to education as a whole is limited. A recent report by the World Bank (2008) states, "Liberia spends approximately 12 percent of government expenditure on education, which is lower than low and middle income countries' average of 20 percent" (2). While estimates have varied on the exact expenditure for teacher salaries, a report by the MOE (2007) explained that due to its limited national budget, salaries paid to teachers were as low as $20 per month.

During the period of early reconstruction, the UN provided administrative oversight for the Liberian MOE, including its tasks of accounting for and delivering salaries to government teachers. In many cases, INGOs, such as the International Rescue Committee, attempted to defray some of the costs to the government by providing stipends to teachers at selected government schools. Due to budgetary limitations among INGOs, this practice was limited in scope, leaving the majority of public schools across the country without a consistent way of paying salaries to teachers.

Social Injustice, Low Compensation, and the Suffering of Teachers

The theory of social justice developed by philosopher Johns Rawls (1971) draws attention to the concept of fair distribution of both economic and noneconomic goods. Specifically, his work has generated interest in how unequal access to adequate social resources, such as limited opportunities for education, health care, and housing, results in social and psychological harm (Swenson 1998). Over the past decades, social psychologists building upon Rawls's work have utilized the concepts of distributive and procedural justice to investigate perceived fairness in workplace settings (see, e.g., Tyler 1997). Distributive justice is concerned with the criteria that lead to a fair outcome, such as distribution of pay, promotion, and so on. Procedural justice examines the "how" of a process or policy and is therefore associated with distributive justice (Deutsch 2006). In this study, I consider distributive justice by investigating inter/national policies that result in very low compensation for Liberian teachers and procedural justice by exploring the procedures for distributing compensation that affect teachers' sense of well-being and the education they provide.

To examine compensation in light of distributive and procedural injustice, I was primarily concerned with how teachers themselves described the forces affecting their well-being and ability as teachers. Data from the card-sort component of my interviews with individual teachers were particularly revealing: "Teacher salary" was overwhelmingly reported as the factor most affecting their well-being by the 66 teachers who were interviewed. Through the interviews, I came to understand that the concern voiced as "teacher salary" referred to a larger problem that included the policies and procedures that affect teacher remuneration in Liberia. Specifically, teacher salary was fraught with difficulties of both distributive (what teachers are paid) and procedural (how teachers are paid) processes during Liberia's early reconstruction period.

At the time of this study, teachers were among the lowest paid civil servants in Liberia. Not surprisingly, teacher pay was often referenced in relation to its practical worth. During interviews, teachers described the inadequacy of their salary by relating it to the cost of buying a bag of rice, the staple food of Liberia, a comparison also made in many media articles, ministry meetings, and reports. For example, while explaining his reasons for stating "low salary" as the issue most affecting his well-being, one teacher stated, "Looking at our contract, we are paid, I would say peanuts, because what we earn cannot even buy a bag of rice" (Interview, April 30, 2006). This teacher reported that a bag of rice cost US$23 in 2006. Even if they could afford an entire bag of rice, one bag would be insufficient to feed the teachers and their families for more than several days. The teachers almost always stated this fact to demonstrate how little their salary purchased under current economic conditions.

The procedure by which salaries were paid was commonly observed as problematic by the teachers in this study. The problem was also frequently discussed at educational meetings with MOE officials, INGO meetings, interagency INGO forums, and in the local and international media. Overall, the discussions revolved around three issues: (1) unreliable distribution mechanisms; (2) an uneven salary scale in relation to qualifications; and (3) disproportionately low earnings that make it difficult to afford the basic costs of living. I relate the first two issues to the concept of procedural justice and the third to distributive justice.

The manner by which teacher salaries were distributed and the policies determining who received what salary were inconsistent and posed a serious problem from a procedural justice perspective. In the early postwar period, the MOE continued to pay teachers in the same manner as it had before the war began. "Government teachers," as they are commonly called by Liberians, are teachers who work at public schools and are therefore eligible to earn a civil-servant salary from the centralized Liberian government. The MOE, which pays teachers according to a very dated system, bases its distribution of teacher salary on a list of names of teachers registered as civil servants. The actual paychecks are then distributed and delivered by district or county education officers to teachers throughout Liberia.

However, long years of war have resulted in limited or no resources for the government offices to update the list. As a result, names of working teachers are often left off the list, while the names of teachers who no longer teach (some of whom have left Liberia) remain on the list. This fact is compounded by the legacy of "ghost teachers" added to the list well before the war. As described by Chapman (2002, 13), an international education researcher:

> This [phenomenon of ghost teachers] was illustrated in pre-civil war Liberia. Given the complexity and corruption in the process of getting replacement teachers hired to replace teachers who died or left teaching (new teachers needed 29 official signatures to get on the payroll), headmasters were allowed to appoint temporary substitutes and let them cash the paychecks of the teachers they replaced. Principals quickly realized that they could cash these paychecks and keep the money, without bothering to appoint a replacement teacher.

The mechanisms used to distribute salaries also created problems. Because salaries are distributed in Monrovia, teachers living outside of the capital, particularly those living in rural areas, had two choices. They could travel to Monrovia, incurring transportation, food, and housing costs that sometimes exceeded the salary itself. Alternately, they could have their salary "brokered" by District Education Officers (DEO). In this system of salary distribution, DEOs were authorized to deliver salaries in cash to teachers living outside of Monrovia. However, the teachers I interviewed reported that their "brokered" salaries were frequently delayed or that, too often, their pay never arrived.

The problems with this salary-distribution system were exacerbated by the collapsed state of Liberia's infrastructure. Specifically, unreliable transportation, unsafe roads, limited material resources, and the absence of a national postal system created significant barriers to the execution and monitoring of this system. As a result, DEOs were unable to deliver salaries, and teachers, especially those living in rural areas, were forced to wait months to be paid. As one teacher stated, "Delay, it [salary] is always delayed... delay, delay, delay" (Interview, May 2, 2006).

Compounding this problem, DEOs often charged a "broker's fee" for distributing salaries. The DEOs claimed that charging this fee was necessary to compensate their efforts in collecting and distributing salaries, despite the fact that these tasks are included in their job responsibilities. Therefore, the actual salary that many teachers received was less than the promised amount. The government's continued use of these "brokers" resulted in a pay process that marginalized teachers, making them dependent on brokers who siphoned off part of teachers' salaries.

Another aspect of procedural injustice that the teachers described regarded the allocation of salary based on teacher qualifications. In Liberia, individuals must have a teacher-training certificate from a Liberian teacher-training institute or a bachelor's or master's degree from an institution of higher education to obtain the requisite qualifications to teach. After the war ended in 2003, the MOE revised its policies in the hope of making it easier for teachers to obtain recognized accreditation. However, obtaining accreditation was still difficult because the few operating teacher-training institutes were understaffed and lacked resources. It has been especially difficult for teachers living in rural areas, where the teacher-training institutes were not functioning. Therefore, at the time of my fieldwork, the majority of teachers teaching in rural schools at all levels (primary and secondary) did not hold formal teaching certificates.

Furthermore, trained teachers often could not provide evidence of their qualifications because their paperwork had been lost or destroyed during the war. The outbreak of violence caused many people to flee their homes with little or no time to organize their possessions. Many lost their paper-work while living in hiding, while migrating to different areas, or when their homes were looted or burned. During in-depth interviews, several teachers mentioned their concern that their salary did not reflect their training as measured by their level of teaching and/or years of teaching experience. Of particular note are the trained teachers who felt that their formal training and years of experience were not reflected in their pay. In a survey of 106 teachers, 81 percent reported that their salary does not correspond to their qualifications. The teachers noted that instead of being determined by their qualifications, their salary is instead determined by "who you know," "favoritism," "who you are related to," and "whoever is in charge." The teachers were referring to the problems of corruption and misuse of power among DEOs, County Education Officers (CEOs), and MOE staff. Several teachers acknowledged that certain teachers received assistance in obtaining their salaries from relatives or friends who worked for the government.

Many teachers expressed feeling traumatized by their poor salary and consequent inabilities to meet their basic needs. While interviewing and surveying the teachers to further investigate how low salaries affect their well-being, I found that teachers' responses primarily revolved around how their pay fails to provide adequate food, shelter, and medical attention. The teachers were unable to provide for their own and their family's basic needs by working between four and six hours a day to earn a salary that for some teachers, after paying broker's fees, were as low as $15 per month. One teacher living in rural Nimba reported, "The salary is very little in that it can't even stay for two days" (Interview, June 13, 2006).

As this teacher stated, their money can hardly "stay" given the costs of food and shelter. Most of the interviewed teachers were married (80 percent) and had children (95 percent). As adult caregivers, they were providers for their immediate and extended families. On average, the teachers interviewed lived in households of eight people; their salaries could not feed the many people dependent on them. Further, the teachers could not afford other basic needs, such as housing, health care, and school fees for their children.

Overall, teachers related a low salary to feeling badly about themselves because they were unable to provide for their families. For instance, 75 percent of the teachers surveyed reported that their low salary also affects their self-esteem. The interview data also revealed that the teachers felt demoralized because they did not have enough money to buy "proper clothing" to "look like a teacher should look" (Fieldnotes, March 24 and April 30, 2006). The teachers felt they could not represent themselves well as teachers. "Students make fun of us, they call us names because we don't look good," stated one teacher (Fieldnotes, May 16, 2006). A female teacher stated that she would rather not come to class than appear in the same clothes every day: "It is not right, that as employed teachers we cannot even dress properly" (Fieldnotes, August 7, 2006). Another teacher pointed out the dilemma that teachers face in terms of receiving a low salary while trying to abide by the regulation in the code of conduct that teachers should dress professionally:

> But it's not conducive, frankly, to work under condition[s] where your salary is so small [that] there's not enough for a good living place. Even wearing, some teachers you see, teachers wearing t-shirt, you know. We have an ethic, in the Education Ministry, a dress code. The teacher should be able to dress neatly, every day. But suppose you don't have the means? Your salary is so small, I will [have] to wear the [same] clothes and shoes [every day]. (Interview, March 28, 2006)

Another serious compensation issue described by teachers was the lack of housing provided to them in the post-war period, even though it had been provided before the war in some of the rural sites. This problem, they explained, increased the instability of their home lives. The war destroyed the majority of houses in all three sites in this study. Teachers who had been displaced and had now returned home had to find new housing or to rent refurbished or intact houses. In addition, the widespread presence of international organizations was driving up the costs of housing in all three counties. Underpaid teachers could not afford the costs of supplies for building new homes, and they were often seen as undesirable tenants because their salaries were known to be low and unreliable.

The poor monetary support and the housing challenges led teachers to become physically weakened and psychologically demoralized. Injustices in training and material support also compromised the well-being of teachers. For instance, teachers were frustrated by the varied competencies of their fellow teachers. Trained teachers reported feeling shame and embarrassment because parents associated them with teachers who had not been trained and were considered incompetent at their jobs. Those who had not been professionally trained and were volunteers were also embarrassed by their limited training background and sometimes frustrated by the procedural policies surrounding certification. Specifically, they were frustrated that hard work was not recognized as equally worthy of certification and, therefore, of government pay. The distributive aspects of classroom materials led to physical, psychological, and economic concerns as well. Without a sufficient number of desks, for example, teachers were forced to stand for long periods, and chalkboard dust from old chalkboards led to eye pain and disease. Furthermore, because they lacked curriculum and textbooks that should have been distributed by the government, some teachers used their own money to buy texts, which meant they had even less money to spend on their basic family needs.

In addition to these material concerns, teachers discussed the challenge of managing their classroom because of changes in their students. The war had serious psychological and physical costs for teachers at their jobs because, among other things, many of their students had witnessed, experienced, or inflicted violence themselves. The unprecedented violence during the war included the widespread manipulation of children into fighting forces and sexual slavery. Other forms of sexual, physical, and emotional violence have also been documented, with gender-based violence especially rampant. Large numbers of children who had been victims of such violence had re-enrolled in schools as part of the Liberian plan for recovery. Thus, classrooms were more crowded, and children carried with them their experiences of years of violence. Teachers frequently reported having trouble "managing" these children's needs. Teachers observed that children were engaged in "bad behaviors" learned in the war, such as substance abuse. Teachers were eager to be trained to help their students. However, these programs were not offered consistently, and they varied in the quality of the material covered, again raising distributive and procedural justice concerns. My findings suggested that teachers wished to receive more training on how to work with their students but could not because the programs were offered in a piecemeal fashion. The challenges to classroom management led many teachers to become worried and to feel frustrated that they were ill prepared to handle the complex needs of their students.

Overall, low salary, poor housing, and challenging classroom management issues affected the way teachers perceived the respect they were granted. On the one hand, more than 90 percent of the teachers interviewed stated that they felt respected by parents and their community for being teachers. One teacher from Monrovia, for example, stated, "And one thing is true, the profession in Liberia, well there is no money but the people in the community respect you a lot, and the children they always want to be around you and you're always around them" (Fieldnotes, May 2, 2006). However, almost all of the teachers interviewed stated that they felt that the teaching profession is not respected at the national level in Liberia. The differences in the level of respect they were accorded by their community and by the government were articulated by an experienced teacher in rural Lofa County:

> The community regards us very highly knowing that they are taking care of their children...but [in terms of national respect] no teacher has that good

feeling. Teaching as our career now, the younger ones are leaving the teaching field because the salary is not impressive. (Interview, March 23, 2006)

This teacher, like others, held the view that their low salaries demonstrated a lack of respect by the national government. "Just look at our low salaries," one teacher argued, "does that look like respect?" (Fieldnotes, April 30, 2006). For this reason, a teacher in Monrovia explained why teachers often leave the profession: "Let's focus on one point. Teachers are people who work very hard and at the end what they earn cannot sustain and serve food for their family. The job they do, it is causing some to not specialize in teaching" (Interview, March 23, 2006).

The phrase "if you want to die poor, be a teacher" was often (re)stated in response to questions about the status of teachers in Liberia today. When I asked teachers how this phrase made them feel, they responded much like a teacher from Monrovia, who answered, "We feel bad, really bad" (Fieldnotes, March 14, 2006). Some had hope that the new Sirleaf administration would bring positive changes. As one male teacher from Lofa commented: "We are looked down upon. The Liberian government did not care about teachers in the country. I do not know why they are not respected, teacher salary is small... still we are looking to this new government for this change" (Interview, March 23, 2006).

Low salaries contributed to teachers' concerns for the country's peaceful development. The combination of inadequate pay, limited teaching materials, and insufficient training led to a general fear among teachers for the future of the teaching profession. They pointed out that not only was there a dearth of teachers because of the war, but those who were qualified to teach were leaving the field for better paying jobs, resulting in an intensification of the existing brain drain. The lack of material support for teachers appeared to fuel mistrust and a lack of confidence in government efforts to effectively redevelop Liberia's education system. In discussion groups, principals, teacher trainers, university faculty, and ministry members expressed concern for the future of the teaching profession and hence education, if certain standards were not met. As a female teacher from Bong County asked, "Who will [be] our future teachers if we don't show the professional respect for the job?" (Interview, June 14, 2006). Similarly, a principal noted:

Most quality teaching in the school is no more what it used to be...Now it is hard because teachers are paid very, very low compared with especially NGOs that are working now. And as a result, most of our trained teachers have left the teaching field. So, people we see today, they are using education, teaching as a stepping stone for other areas. That's one of the big differences we see. And maybe that's also, that will also contribute to the behavior of children. Because children are not given the right teaching...Many of our trained teachers have left the classroom because a teacher who is teaching cannot get money to buy a bag of rice who has a family. So this has discouraged many teachers. (Fieldnotes, March 23, 2006)

Teachers suggested that the precarious future of the teaching profession was a direct result of poor salaries, and they considered this problem a threat to the security of Liberia. This issue was discussed at length in a focus group with women, who explained that some schools were demanding "school fees" from parents and the community to supplement teachers' salaries. Only the wealthier, urban families could afford schooling under these conditions. This practice attracted the most qualified teachers to urban schools, leaving people in rural areas more vulnerable to poor quality education. This problem was powerfully described in the following statement by a rural teacher during a focus group discussion: "It is the feeling all over again, like the war, a few get education, the most don't." She continued:

At the University, well, they [the Ministry] only want a few people there educated, only special ones; [they do not want] everyone educated. The majority of us are not educated, very few of us [teachers] will go to the University. The government downplays us, we don't want to suffer all of the time. We [rural teachers] want PhD too, to be educated, to live a good life with people, but without education, there is nothing, no peace...The Ministry knows this, they are aware, but they downplay us teachers. (Focus Group Discussion, August 15, 2006)

Thus, the lack of support for teachers not only affects their well-being and the quality of their teaching, but, they argue, it also threatens Liberia's future for peaceful development. This was a view shared by some MOE officials, who also feared that the dearth of teachers, particularly in rural regions, could reignite Liberia's old problems of differential access to education between rural and urban residents.

Concluding Thoughts on Supporting Teachers and Liberia's Peaceful Development

The situation of teachers in Liberia's period of early reconstruction has considerable implications for the peace of the nation and the well-being of its teaching force and students. This case also has implications for education planning in fragile states because the problem of teacher compensation is not limited to post-war Liberia. International policies provide limited support for the MOE to provide teacher salaries, thereby negatively affecting the development of a committed, quality teaching corps necessary for the revitalization of Liberia. When salary delivery structures are as limited after a conflict as they are in Liberia, it is not realistic to give the task to a ministry already understaffed and underprepared for the multiple educational challenges it faces.

Similar to the chapters by Muro Phillips (see chapter 3) and Valdiviezo (see chapter 8), the vertical case study approach helped me to explore how local teachers make sense of, embrace, and resist their positions in light of the inter/national influences with which they must interact. In these chapters, important implications about the overall health of teachers and students and the participation of teachers and community members in policy appropriation were examined in relation to the globalizing forces that shape their daily lives.

In line with Mendenhall's and Zakharia's discussions of post-conflict education in this section, my research also examined the friction produced by continuous social interaction among policy actors at different levels. My particular focus has been, on the one hand, the unstable connections between international and national teacher compensation policies and, on the other, the interaction between national policy and the teachers themselves. This useful friction metaphor helped me to consider, in the Liberian case, "the awkward, unequal, unstable, and creative qualities of interconnection across difference" that "inflect historical trajectories, enabling, excluding, and particularizing" (Tsing 2005, 6). The findings from my research prompted attention to these connections; in this particular case, the instability perpetuated by these interconnections has the potential to ignite historical tensions in this fragile state. The importance of providing material and psychosocial support to teachers, especially in post-war situations like Liberia, cannot be overstated. Restoring a sense of justice among the teaching force is an important first step toward rebuilding the entire country.

Notes

1. Similar to Mendenhall (chapter 10), I define "fragile states" as those likely to experience conflict or recently emerging from a domestic or international conflict and those with very weak institutional structures to avert future conflict (Crisis States Workshop 2006).

2. A focus group discussion was held with female teachers because survey and interview data prompted my further interest in how female teachers, in particular, experienced violence, corruption, and limited access to teacher training.

3. See, e.g., the *Master Plan for Education Liberia: 2004–2015* and the *Girls Education Policy* (2006).

Chapter 12

Positioning Arabic in Schools
Language Policy, National Identity, and Development in Contemporary Lebanon

Zeena Zakharia

Scholarship on language and society points to language as a significant site for ideological contestation and identity assertion (Suleiman 1996). Contemporary language policy debates in the Middle East are no exception, and they are intimately shaped by broader sociopolitical and economic processes that affect the region. In Lebanon, language policy in education has reflected discourses about national identity and development, implicating formal schooling in religious inequality and sectarian struggle over the last century. Post-civil war peace accords, brought about through the mediation of Arab regional actors in 1989, sought to unify a divided country, in part, by officially asserting Lebanon's Arab identity and affiliation through constitutional amendments[1] and educational reforms. The resultant language policy in education, launched in 1997, undertook the reinstatement of the Arabic language as the common language for all students in Lebanese schools.

This chapter looks at the positioning of the Arabic language and its (non-) speakers in contemporary national education policy and in local schooling practices across different religious, socioeconomic, and language-schooling contexts in Greater Beirut. In particular, it considers how the Arabic language, made central to post-civil war national unity, is simultaneously promoted and undermined by school, state, regional, and global actors during periods of regional and national instability and violence.

To understand this concomitant cherishing and devaluing of the Arabic language in contemporary Lebanese educational discourse and practice, this chapter examines complex interrelated units of analysis vertically and horizontally across time and space. Through an "ethnography of global connections," I explore how the friction within and between local and global actors has generated new cultural and political forms that influence school-based language policy and practice (Tsing 2005, 1). I focus particular attention on domestic (Lebanese) and regional (Middle East) conflict in the context of the global political economy and how it affects educational policy formation and identity assertion among Lebanese youth.

I begin with a brief historical overview of language issues in Lebanon and the region to contextualize post-civil war national language policy, which sought to reinstate the Arabic language to a prominent position in light of colonial and missionary legacies in education. Drawing on the notion of policy appropriation (Sutton and Levinson 2001; see chapter 1, this volume) and language as a site for ideological struggle and identity assertion (Suleiman 2004), I examine how political and economic events have positioned the Arabic language over time. I then analyze the positioning of the Arabic language within the context of political instability, violence, and economic uncertainty that characterized the field research period between 2005 and 2007. I argue that Arabic, though hailed as the language of collective national identity, is undermined by a variety of factors. These include economic pressures and concerns about personal security that promote the study of French and English; the stigmatization of monolingual Arabic speakers as not modern and schools with widespread use of Arabic as deficient; and mundane school practices, including implementations of the national curriculum, which place lower priority on learning Arabic.

Collective Identity and the Making and Remaking of Language Policy: A Historical Overview

The Arabic language has enduring significance for the conceptualization of secular and nonsecular collective identity in the Middle East (Suleiman 1996, 2003). During the precolonial period, the spread of Arabic followed the expansion of Islam and the scholarly and educational works of the religious elite, who played a vital role in teaching literacy throughout the urban centers of the region (Holt 1996). The survival of the standard variety of Arabic from its Qur'anic roots, the development of print capitalism, and corresponding technological innovations in the first half of the nineteenth century allowed for the spread of ideas to a diverse readership in various regions of the Arabic-speaking world and beyond (Holt 1996). This expansion enhanced the bond across a region differentiated dialectically, geographically, and religiously—a process that would continue after the dissolution of the Ottoman Empire and independence from European colonization (Suleiman 1996). Lebanese Christians, a minority in the region, contributed to the promotion of the Arabic language as the basis for modern national identity and to unify a diverse populace in a mounting resistance against Ottoman rule and Turkification (Suleiman 2004). Thus, Arabic was central to the symbolic contestation of Turkish Islamic imperialism; it became the embodiment of affiliation among speakers across the region, regardless of their religion.

The introduction of Western languages to formal education in Lebanon is attributed to European and American missionaries during the late Ottoman period (Shaaban and Ghaith 1999). Fueled by competing interests in the region, French Catholic and various British and American Protestant missions established bilingual schools alongside their Greek and Russian Orthodox predecessors and the modern Turkish-Islamic Ottoman schools. By the late nineteenth century, educational ties were established between various religious communities and their foreign missionary sponsors (Frayha 2004), with the exception of the Shi'a, who were not engaged in the educational enterprise of the period (Abouchedid and Nasser 2000). Thus, through a process of differentiation, contemporaneous with the emergence of sectarianism as both a discourse and practice in nineteenth-century Ottoman Lebanon (Makdisi 2000), particular languages came to be associated with particular religious groups in Lebanon.

Although some in the Arabic-speaking world interpreted the introduction of nonnative languages into schooling and other areas of social life as "cultural invasion" (Suleiman 2004, 27), others actively sought the establishment of foreign Christian missionary schools in their communities as a means for educational advancement. At the same time, by the late 1800s, various Lebanese Christian and Muslim school systems emerged as "national" alternatives to European missionary schools. These schools, together with their foreign counterparts, constituted language experiments in education, promoting Arabic and foreign languages through their practices.

Fears about a linguistic invasion of Arabic in Lebanon relate to the perceived subordination of the Arabic language and Arab culture and scholarship to the West (Suleiman 2004). This concern dates back at least to the missionary educational enterprise, during which Arabic was assigned a literary function in education while foreign languages were assigned a scientific and modernizing function. The French, in particular, actively sought to spread the French language among Catholics and Maronites. Under French rule from 1920 to 1943, French was introduced alongside Arabic as an official national language in the newly established constitution of 1926.

Although Arabic was made the sole official language at independence in 1943, French remained prominent in private institutions and public life (Massialas and Jarrar 1991). In addition, in 1946, the state made English an official alternative to French in the bilingual system of schooling. However, influenced by Arabization movements in education, which saw the use of French and English as "expressions of 'cultural colonisation'" (Frayha 2004, 173), government-initiated curricular reform in the late 1960s pushed to make Arabic more prominent as the medium of instruction in public and private schools to foster a shared national identity.

The civil war (1975–1990) interrupted efforts to re-establish Arabic as the language of instruction for all subjects other than foreign languages and literature. With conflict and state failure came the demise of public schools and the proliferation of private institutions, which continued to promote French and English. As with earlier Ottoman language patterns, French resumed its association with the educated Christian elite, and English came to be associated with educated Muslims or Orthodox Christians (Joseph 2004). By the 1990s, 75 percent of French-speaking Lebanese were Christians (Abou, Kasparian, and Haddad 1996).

Thus, when post-civil war constitutional amendments made way for educational reforms in 1990, language policy was seen to be one of the "critical and sensitive issues...particularly concerning the position of the Arabic language" (Frayha 2004, 185). Launched in 1997, the current national curriculum requires Arabic for all students, linking Arabic language learning to an identity-related discourse (Zakharia 2009). In addition, either French or English serves as the second medium of instruction starting in the elementary grades, and a third language (French or English) is studied starting in the seventh grade (NCERD 1995). In keeping with earlier educational practices, humanities and social science subjects are taught in Arabic, and mathematics and sciences are taught in English or French, according to the language tradition of the school (Zakharia 2009). Today, approximately 70 percent of schools in Lebanon teach in Arabic and French, with English as a third language. In Beirut, during the 2006–2007 academic year, 43 percent of schools taught in Arabic and French, 30 percent taught in Arabic and English, and 27 percent of schools taught both Arabic-French and Arabic-English programs (CRDP 2007).

Conceptualizing a Study of Post-Colonial and Post-War Language Policy in Education

Language is a complex site for ideological contestation and identity assertion, where asymmetrical power relations exist between groups and individuals (Suleiman 2004). As Suleiman describes, "Language always stands at the crossroads of (social) time, linking the past with the present and linking these two with the future. This is particularly poignant, both politically and culturally, for those languages with a long recorded heritage, of which Arabic is an example" (2004, 7). Such linkages are also poignant for their speakers during periods of instability and sociopolitical conflict. In Lebanon, nationalist ideologies and state-centered discourses have endeavored to promote and circumscribe national identity through school-based instruction in a common language: Arabic. However, this facile ideology obscures the complex ways in which children and youth are socialized into language beliefs and practices in multiple spaces, including their schools, homes, neighborhoods, and community organizations (García and Zakharia forthcoming)—a socialization process influenced by social, political, and economic circumstances, including vulnerability and threat.

Research on language and nationalism focuses on the state as an intentional actor, who through centralized strategies, attempts to impose policies and shape national identity. In reality, the state is not a unified actor, but rather functions as an assemblage of policy actors, including those at the substate and superstate levels who actively appropriate policy toward their own ends (Blommaert 2006). The friction produced by the continuous social interaction among actors at these various levels with diverse social and political goals often constrains the actions of the state and generates local appropriations of national policy (Latour 2005; Tsing 2005; see chapters 1 and 2, this volume). Thus, this study focuses not on the state's strategies to impose an Arabic language policy but rather on the contradictory operations of policy at the more localized sites of schooling. By treating language policy as an interactional, interpretive process (Sutton and Levinson 2001; Yanow 2000), I explore a range of sites for understanding language policies as spoken, written, and enacted, tracing the interrelatedness of complex global, regional, national, and locally specific phenomena.

This vertical case study, therefore, considers how language policy in education is socioculturally and politically produced, or constructed and reconstructed by an assemblage of interconnected actors, through a process of appropriation that is informed by political and economic events that are simultaneously local and global in relation to both time and space. Drawing on multiple methods and sources of data collected during 21 months of fieldwork conducted in Greater Beirut between May 2005 and October 2007, the project involved 3 interdependent levels of investigation. First, at the broadest level, I conducted an inquiry at 10 schools.[2] There, I surveyed 1,000 secondary school students to generate cross-school self-report data on the language beliefs and practices of students; I also interviewed selected school administrators and teachers and conducted school-based observations. Second, I undertook a multisited ethnography of three of the ten schools (one secular-Arabic-English; one Shi'i-Arabic-French-English; and one Sunni-Arabic-French-English), focusing in greater depth on how schools appropriate national policies. Methods used included participant observation; classroom visits; interviews with school administrators, teachers, and students; focus groups with students and parents; and document analysis. The ethnography also involved conversations with school network administrators and Ministry of Education officials. The third level of investigation engaged six focal students—two from each focal school—in in-depth interviews and observations to understand the lived linguistic experiences of students and the "acting out" of their self-reported beliefs and practices about language and identity.

While the study began as an investigation of post-colonial and post-civil war language-in-education policy, during the period of field research, a war between Israel and Hizbullah in Lebanon devastated the country, displacing one million people, damaging more than 350 schools (Shehab 2006), and drastically changing internal alliances between political leaders and their supporters. The 34-day war, which began in July 2006 and disproportionately targeted Shi'i towns and neighborhoods in Greater Beirut, was followed by internal sectarian violence, protests, sit-ins, and riots. A series of politically motivated assassinations, roadside bombs, and other violent events ensued and resulted in daily disruptions to schooling. By 2007, a second war between the Lebanese army and an al Qaeda-inspired Sunni insurgent militia was underway in northern Lebanon. Several of the schools in this study sheltered orphans and displaced youth, including Iraqi refugees from the U.S.-led war in Iraq. Thus, violence, political instability, and concerns over security permeated the research context, creating new vulnerabilities for teachers and students and heightened awareness of social (in) justice (see chapter 11). In the next section, I examine the positioning of the Arabic language in schools in light of this renewal of tensions.

Positioning the Arabic Language in Schools

Despite a mandated national curriculum, schools and school actors in Beirut's schools engage with different language policies and language realities in their schools, classrooms, and neighborhoods. Any discussion of language policy or practice in schools evokes comparison between languages and schools and their students. As a Shi'i youth at the Sunni School[3] commented on her survey, "There is a big difference in the way of teaching languages from one school to another; moreover, the usage of foreign languages depends on the region you are in" (2007). Thus, youth and other school actors are aware of the differences in linguistic resources between schools.

Despite these differences, however, the study revealed that youth and their teachers cherish the Arabic language while, in seeming contradiction, view its use on school campuses as a deficit. The majority of students across the 10 participating schools, regardless of stated religious affiliation, expressed a strong connection to the Arabic language, constituting a shift from pre-civil war patterns (Zakharia 2009). According to survey findings, for example, 86 percent of the 1,000 secondary school participants agreed that Arabic was important in their lives. This finding was supported by interviews and focus groups with Arabic speakers and learners at the focal schools. As a 16-year-old Maronite youth at the Arabic-French Catholic Brothers School wrote: "We cannot let go of the Arabic language no matter what, because it is proof of our identity." Lebanese youth who had lived abroad for a large portion of their lives, and therefore did not speak Arabic well, also expressed a strong connection to the Arabic language. For example, Mark, a Lebanese-American 16 year-old at the Arabic-English Protestant School, explained to me:

> I feel a strong connection to the Arabic language even though I only started learning it when we moved to Lebanon from the US three years ago. I am working hard to learn it because I want to make sure that I will be able to pass it on to my children in the future. That is really important to me because it is part of who I am. I want to maintain my culture. (Interview, in English, June 13, 2005)

As with Mark, the high value placed on the Arabic language was frequently linked to a cultural or collective identity narrative for both Arabic speakers and nonspeakers of Lebanese descent. In fact, 87 percent of students responded that losing Arabic would constitute losing part of their cultural heritage. According to Ali, a Shi'i youth at the Arabic-English Protestant School:

> As a child, I did not like the Arabic language; I used to hate Arabic.... But when I grew up, I realized that the Arabic language embodies a person's potential. It secures one's *ḥaḍara* [civilization and culture or grand traditions]. It gives one strength for his country and connects him to the nation.

I consider that whoever gives up his language is abandoning his *haḍara* and leaving everything behind. Lebanon without Arabic is not Lebanon.... The "major major major" indicator of affiliation to the nation is the Arabic language. Arabic is the nation. Arabic is nationality. (Interview, in Arabic, June 14, 2005)

By relating the Arabic language to "potential" and to *haḍara*, Ali was linking himself as an Arabic speaker to the future and to the past through the Arabic language which, in his view, embodies the nation and unites diverse people.

Similarly, school administrators engaged the Arabic language in various identity-related narratives. For example, an administrator at the Secular Arabic-English School, in explaining that grade reports and communications with parents are conducted in Arabic, said:

> The general outlook is that we are an Arab heritage school, unlike other [comparable] schools...So that's why we choose to present ourselves in Arabic...We teach the ethics and values of Lebanon and the Arab world, but prepare students for success anywhere...We are unique that way. This bond [with our Lebanese and Arab heritage] has always been the differentiating factor [between other schools and ours]. (Interview, in English, October 17, 2007)

Such views suggest pride in the use of Arabic connected to an Arab identity.

However, at the same bilingual school, which is well-known for high success rates on the Lebanese Baccalaureate and a strong university matriculation record, another school administrator explained that the prevalent use of Arabic reflected a deficiency; students and teachers should be using English, but they communicate largely in Arabic in and out of class:

> You mainly hear Arabic in the hallways. I would say that 80 percent of what happens in this school is in Arabic, and English happens in English classes only. Most of our teachers have not mastered English, very few, maybe 10 to 15 percent. This is a school that is based in English, and we should do things in English, but we don't; we do everything in Arabic. Our circulars and general staff meetings are in Arabic; teachers ask for this. I am not saying that Arabic is not important; I am Lebanese and an Arab first. (Interview, in English, October 16, 2007)

The administrator thus indicated that Arabic is important for Lebanese and Arab identity but its widespread use on campus was not desirable. This view resonated with the widely held idea that "good" schools, or schools providing a quality education in Lebanon, demonstrate a high level of foreign language competence, and operate in that language.

Administrators and teachers at various schools expressed similar concerns about the widespread use of Arabic in and out of class. Expectations and unwritten language policies required teachers to deliver mathematics and sciences in English or French, and to teach humanities and social sciences in Arabic, but this policy was differentially applied across schools. Furthermore, teachers and administrators frequently blamed the perceived failure of their foreign language programs and policies on youth and their families, linking the use of Arabic on campus to deficiencies in students' bilingualism. According to the administrator at the Secular School, for example:

> We were obliged to take students who are sub-par this year...The students speak in Arabic in the playground. It is a cultural problem. They are not exposed socially to societies that use these [foreign] languages. They have not traveled. Even their maid speaks better English and teaches the children. (Interview, in English, October 16, 2007)

The administrator then explained to me that 60 to 70 percent of students at the school were now coming from Dahiyeh,[4] Beirut's southern suburb, even though the school is situated in a Sunni-majority neighborhood. Thus she signaled that, in her view, the students with a "cultural problem" are Shi'i Muslims. Other school administrators and teachers also referred to changing student demographics to explain the wide-scale use of Arabic by students in the hallways and playgrounds, or what they perceived to be a decrease in English or French language use at their schools. At a participating Catholic Arabic-French School, for example, an administrator explained, "Here we speak French with students wherever possible, but in the playground, they speak in Arabic." He then went on to say:

> The Shi'a have become the smartest people; they are putting their children in the best schools [like ours]. There is a change in our school populations. We are Maronite and nationalist at the same time; we do not differentiate between Muslim and Christian students. But these families, they don't speak to their children in French at home, and that's why they don't speak French at school...The Muslims say they want to send their children to our schools to learn the Christian way, for a better education...We are losing the cultured students...We have a few of these families left, but not like before. We have mostly working-class people now. (Interview, in Arabic, October 16, 2007)

The administrator thus related the prevalence of Arabic on campus to deficiency in students' and their families' bilingual competence, and further,

to a change in student demographics, namely an increase in Shi'i enrollments. Similarly, in reflecting on the language practices at his school, the
head of a low tuition school within the same network of Catholic schools
told me, "Students from poor families lower our standards, but it's important to have them in schools, even though we lose the cultured parents who
don't want their children in school with the others... I am doing my best
to change the demography" (October 19, 2007). He then clarified that he
has been developing policies over the past five years to decrease the number of Shi'i students at the school, currently 17 percent of the population,
because they "create problems," presumably because they are not from
the same kind of "cultured" families as other students. Like the Arabic-
English Secular School and the Arabic-English Evangelical School in this
study, the interviews with these school administrators suggest that socioeconomic and religious shifts are underway in schools with long-standing
reputations, as parents of historically marginalized communities seek better opportunities for their children's schooling.

Not all schools perceived the increased heterogeneity of their schools as a
problem; nevertheless, during the period of research, certain types of diversity posed a challenge for schools seeking to maintain a politics-free zone for
students amidst rising tensions between various groups. These demographic
shifts became more salient after the July 2006 war, with internal displace-
ments and rising tensions between student populations. They also point to
how a perceived decrease in the bilingualism of schools was attributed to
the increase in Shi'i enrollment or in the number of Muslim students from
lower socioeconomic backgrounds more broadly. There was a stigma associated with being monolingual in Arabic because students and teachers considered it a sign that the individual was not modern or cultured. For example,
when I asked a student at the Shi'i school whether students heard foreign languages spoken in their neighborhoods, she laughed and said, "No way! In my
neighborhood? They are so backward" (Interview, in Arabic, June 2, 2007).
In another case, a 16-year-old Druze male at the Arabic-French Catholic
Brothers School noted in the open-ended portion of his survey: "It is nice for
the Lebanese to know his main language, which is Arabic, but he should also
be cultured and know several different languages" (2007). Such sentiments
were echoed by other students who commonly expressed a multilingual ideology, regardless of whether they were multilingual themselves.

Although they criticized Arabic monolingualism, students, teachers,
and administrators expressed a high value for Arabic, regardless of religious affiliation and school attended; however, they also conveyed a widely
held belief that *other* Lebanese people and their schools do not value the
Arabic language. For example, a female Sunni student in the English section at the Sunni School said, "Education in Lebanon differs from one

school to another. Some schools consider Arabic to be important and some don't care about its existence" (Survey 2007). Similarly, a student from the French section of the same school commented: "Certain groups that belong to a particular and specific religion prefer talking and communicating in a language different than the Arabic language" (Survey 2007). Although neither of these youths specified which groups were perceived not to care about Arabic, the explanation was spelled out to me by Farah, a Sunni student who told me that, in her view, Christians were less likely to value the Arabic language (Interview, May 7, 2007).

Conversely, Arabic language teachers at Christian schools expressed the belief that students at Muslim schools were more likely to take the language seriously and promote it through their educational policies because of the connection between Arabic and Islam. However, teachers at the Islamic schools reported—and I observed—frequent misbehavior in Arabic classes. Thus, while the personal connection to the Arabic language appeared to be strong across all the schools, this did not mean that students behaved well in their Arabic classes. Administrators, teachers, and students complained about student misbehavior in Arabic classes almost across the board. Teachers attributed students' behavior to their indifference to the learning of Arabic, and they blamed the students and their parents for not having a sufficient appreciation of the importance of the language for Muslims and for national culture. For example, Hala, a monolingual Arabic language teacher with 32 years of teaching experience, explained:

> If students were more connected to their religion, they would have a deeper connection to the [Arabic] language. But at this age, they are distant from religion. . . . And the parents, they don't have time, or they are not educated, or simply ignorant. . . . Their national consciousness is no good from the basis. They have zero culture—they and their parents. They make fun of the literary and cultural giants of our times by asking silly questions. They don't give the language value and they don't value their teacher either. (Interview, in Arabic, February 13, 2007)

Thus, Hala associated students' irreverence toward the study of the Arabic language with both their lack of religiosity, or relation to Islam, and their lack of national consciousness. In addition, the perceived ignorance of parents who have "zero culture" compounds the problem, in her view. Other teachers at the Arabic-English Protestant School blamed student misbehavior and seeming disinterest in Arabic classes to the lack of interest on the part of parents, who do not come to the office hours of Arabic teachers the way they do to inquire about their children's progress in mathematics and sciences. Yet Mariam, an Arabic teacher at the Shi'i School, gave a political explanation for this phenomenon of misbehavior, in line with the discourse of the government opposition movement: "A government that does not care for the mother tongue will have a negative impact on the mother tongue" (Interview, May 3, 2007). She went on to explain how aspects of the national curriculum are particularly problematic for promoting the study of Arabic:

> The curriculum creates inordinate pressures on students by making them study mathematics and sciences in a foreign language. This impacts the mother tongue. By Secondary 1 students lose the opportunity to advance in Arabic. The number of hours allocated by the curriculum is severely reduced to make way for the study of sciences, and the program no longer includes any study of grammar. . . . In addition, the government textbooks for Arabic are severely lacking in content. This does not promote the development of an Arabic speaker. (Interview, in Arabic, May 3, 2007)

In interviews and focus group discussions with students from the Arabic classes I observed, I asked students to explain the discrepancy between their stated value for the Arabic language and classroom behavior. Some students cited boredom arising from "boring" or "traditional" teaching methods, literary texts, and topics as the reasons for their misbehavior. Other students cited recurrent topics about peasant life, for example, which they thought were irrelevant to their lives. Most notably, however, students identified the "coefficient," or the system of weighting grades for different subjects, as the reason for not taking Arabic seriously. As one student explained:

> [Arabic] does not have many grades on it, that's why [we don't take classes seriously]; just because of that, not because of the language, but because of the grades. For example, math has more grades on it and later, for university, for example, if we want to take entrance examinations, we have math. It's true that we have an English and math exam, but ultimately, we really need to know our math and science subjects if we want to do well. (Interview, in English, May 7, 2007)

In referring to "grades," this student was talking about the school policy of assigning to subjects relative weights that have been established by the government for the Lebanese Baccalaureate examination. Arabic receives a low weighting on the Baccalaureate for students in the sciences streams. Where schools apply the government coefficient scheme to determine students' school grades and promotion criteria, subjects with higher weights, namely, mathematics and sciences, are privileged. Thus, students, who are under tremendous pressures to be promoted to the next grade level, will focus their study on the subjects that are likely to give them the strongest advantage, that is, those subjects with the highest coefficients: "We have to cut our losses, Miss. With all these subjects, we cannot study for every-thing. We need a period that we can take a break in too. It happens to be Arabic" (Interview, in English, April 30, 2007).[5]

The government policy of assigning relative weights to subjects for the Baccalaureate, therefore, has had the unintended consequence of under-mining the status of the Arabic language in schools. Because they ulti-mately prepare students for this high-stakes examination, many schools also apply the government coefficients to their internal promotion criteria for students to move from one grade to the next. By applying these guide-lines, schools also undermine the status of the Arabic language in the eyes of students, parents, and teachers alike. Schools like the Arabic-English Protestant School, which do not follow this scheme and apply their own coefficients for internal promotion, are taking steps to address the asym-metry between the Arabic language and other major subjects by weight-ing them equally. Nevertheless, these schools also suffer from disciplinary issues. Thus, the misbehavior in Arabic classes across the schools in this study may result from a combination of issues: the national coefficient policy as it is enacted at the school level, the pedagogy used in teaching the Arabic language, and the content of the classes themselves.

Further, students generally sensed that Arabic was not valued because the literature and humanities section—the only track of the Lebanese Baccalaureate in which a high coefficient for the Arabic language is set in the secondary grades and for which several subjects are taught and examined in Arabic—had been eliminated at nine of the ten participat-ing schools. The elimination of this section of the Baccalaureate in many schools was seen by some students and teachers as a blow to the Arabic lan-guage. As the only stream of the Baccalaureate that sets a high coefficient for the Arabic language in the secondary grades and for whom several sub-jects are taught in Arabic, the program's elimination seemed to students, teachers, and school staff as devaluing the language.

In sum, the views expressed earlier suggest that there is a widespread sense that the Arabic language has been devalued in Lebanese secondary schools despite the national policy of promoting Arabic in the education system. Furthermore, contrary to teacher perceptions that the decline of Arabic is due to students' and parents' lack of religiosity, national conscious-ness, or a general devaluing of Arab culture, mundane school practices, including the implementations of the national curriculum, serve to under-mine the status of the Arabic language and student interest in learning it beyond conversational use. In keeping with the discourse of the national curriculum, however, conversations in schools about language and educa-tion frequently linked the Arabic language to a highly valued cultural or national identity, whereas English, French, and other foreign languages were discussed in instrumental terms as languages for communication as "educated" or "modern" persons. A student highlighted the tension created between Arabic and the foreign languages of the curriculum by writing: "It is obvious that in some regions in Lebanon, people like to speak foreign languages more then their own. This is because we Lebanese are being more opened to the outside world which is a very nice result. However, this should not affect our own history and traditions" (Survey 2007).

As this excerpt illustrates, there is a tension surrounding Arabic in relation to the perceived benefits of foreign languages. The political and economic situation compounds this tension between valuing Arabic as Lebanese and/or as Arabs, on the one hand, and knowing that it is not sufficient for future employment in an increasingly mobile world on the other. Students in both Arabic-French and Arabic-English programs were almost unanimous that English would be the most important language for securing their futures. The tension of this realization was exemplified by Ali at the Protestant School, who explained:

> When I grow older, I think I will face problems. I might not be able to find employment.... I will need to know several languages and will not be able to rely solely on my Arabic, nor will I be able to rely on my English, nor my French alone. I must have the goal of having the basics in all these languages. I wish I could do all these things in Arabic, in a language that I love. But not all things that we wish are what will be. I love Arabic, but I know that in the future, when I become an adult, and I need to find work, the opportunity may be elsewhere... I cannot just cling to the Arabic language and say, "I don't want anything except Arabic." (Interview, in Arabic, June 14, 2005)

Another Sunni male studying in the French section of the Sunni School, in commenting about the devaluing of the Arabic language, drew connections to economic pressures and concern about securing a future, saying (in Arabic):

> Schools are not at fault in any way; they strive to teach other languages as well as the mother tongue. However, the fault is that companies, or some companies, in Lebanon don't give any importance to any person holding any kind of diploma or having any specialty if the applicant does not have knowledge of a foreign language. This practice weakens our attachment to the Arabic language, and this is where the mistake lies. (Survey 2007)

The power of this statement in implicating the economy, coupled with the perception of many other students regarding English opening doors to universities, points to the ways in which youth seek to reduce their economic vulnerability through language learning. In privileging foreign languages, namely English, the economy creates a push toward learning foreign languages that may compromise their interest in studying Arabic with equal vigor. At the same time, the vulnerability created by sociopolitical conflict in the Middle East, and in Lebanon in particular, creates a pull toward Arabic to demonstrate patriotic ties. Thus, the language choices of youth and school choices of parents cannot be understood without recognizing the role that internal and regional conflict and the political economy plays in the process.

Concluding Thoughts: Language and Vulnerability

Though promoted through national educational policies, the Arabic language is undermined by various actors through a policy process that alternately positions Arabic as first cherished and then devalued vis-à-vis other school languages and subjects. The national coefficient policy, national testing policies, national and school-level curriculum, and classroom-level pedagogy all contribute to this ambivalent positioning. Teachers and administrators frequently employed a deficit discourse about the monolingual Arabic speaker to discuss demographic changes in their schools and a perceived decrease in bilingualism. Such constructions of difference draw on sectarian analyses, ignoring the impact of policies, curricula, and pedagogy.

Further, it is important to remember that students themselves are critical policy actors. In contrast to commonplace perceptions by teachers and school administrators, secondary school youth across religious, socioeconomic, and language-schooling contexts articulated a strong connection to the Arabic language couched in notions of national identity and patriotism during these turbulent times (Zakharia 2009). However, the vulnerability engendered by the political economy shapes the students' perceptions of which languages are most instrumental for their futures. Policy-related actions are clearly shaped by forces beyond the classroom, and actors negotiate the realities of their linguistic choices in relation to these forces and in tension with their own values. As this analysis of policy as process has shown, broad policy dictates are "charged" by the friction of interaction among actors at multiple levels and "enacted in the sticky materiality of practical encounters" (Tsing 2005, 3). Myriad connections link Lebanese schools and students to regional and global forces that are, or will be, affected by the decisions about language and identity of local policy actors.

By linking language with domestic and regional security concerns, I have sought to further understandings of language policy in education during periods of violence and economic uncertainty. As Shriberg and Mendenhall have also documented in this section, an acute sense of individual vulnerability and of the fragility of the state significantly influence policy actors, such as students, teachers, administrators, and state officials. The friction produced by interactions among these actors at different levels is the result of their different concerns, vulnerabilities, and resources. Thus, the implementation of post-war educational policy by the state can have unintended consequences for actors at other locations. Further, like other chapters in this volume, this study demonstrates how national and school policies are shaped by inter/national pressures and the perceived need to manage diversity and group allegiances through educational practices (see chapter 7). Similar to the analysis in chapter 8, for example, I have highlighted the contestation of official language policies in light of complex sociopolitical, historic, and economic realities and language asymmetries to illustrate how language needs become social justice concerns. In line with the observations in chapter 9, through a process of "othering" by administrators, teachers, and students, the religious identification of youth was made prominent in discussions regarding, in this case, the positioning of the Arabic language, making religion and sect the primary lens through which the "other" is viewed during periods of insecurity.

In Lebanon, constructions of difference were deployed to explain the schools' perceived failure to implement national language policies. Such explanations obscure what is in fact a complex policy process in which youth, as policy actors, are engaged in negotiating their social, political, and economic circumstances, including vulnerability and threat. This chapter supports the argument made by others in this volume that to understand policy processes in education, analyses must engage the significant role that the political economy plays in the process of differentiation. In this way, the vertical approach to comparative and development education makes it possible to understand the contradictory actions of a wide range of policy stakeholders within their social context and within the political and economic realities beyond the classroom that shape their actions and struggles.

Notes

1. In 1990, a Preamble was introduced to the Lebanese Constitution that states, "Lebanon is Arab in its identity and in its affiliation. It is a founding and active

member of the League of Arab States and abides by its pacts and covenants"
(Dabbagh et al. trans. 1997, Preamble no. 2).

2. The 10 schools were selected using sampling categories and a map grid to select
   a diverse and representative sample of Beirut's differentiated neighborhoods.
   Schools included religious (Shi'i, Sunni, Catholic, Evangelical, etc.) and secular and secular mission, tuition-based and fee subsidized (by government or
   religious order), for-profit and non-profit schools, operating in Arabic-French,
   Arabic-English, and Arabic-English-French in 10 distinct neighborhoods of
   Beirut. Stereotypes of the neighborhoods based on religious, socioeconomic,
   political, linguistic, and other associations designated some of the research
   neighborhoods as worldly and hip, and others as Christian, Sunni, or Shi'i
   ghettos.

3. All school and participant names are pseudonyms.

4. Dahiyeh is a diverse region with a Shi'i majority. During the July 2006 war,
   its people suffered massive Israeli aerial bombardment, resulting in large-scale
   displacement.

5. By Secondary 1, or Grade 10, students carry as many as 15 subjects in school.

References

Abou, Sélim, Choghig Kasparian, and Katia Haddad. 1996. *Anatomie de la Francophonie Libanaise.* Beirut: Université St-Joseph; Montreal: AUPELF-UREF.

Abouchedid, Kamal, and Ramzi Nasser. 2000. "The State of History Teaching in Private-Run Confessional Schools in Lebanon: Implications for National Integration." *Mediterranean Journal of Educational Studies* 5 (2): 57–82.

Abowitz, Kathleen Knight, and Jason Harnish. 2006. "Contemporary Discourses of Citizenship." *Review of Educational Research* 76 (4): 653–690.

Abu el-Haj, Thea Renda. 2002. "Contesting the Politics of Culture, Rewriting the Boundaries of Inclusion: Working for Social Justice with Muslim and Arab communities." *Anthropology and Education Quarterly* 33 (3): 308–316.

———. 2007. "'I Was Born Here, but My Home, It's Not Here': Educating for Democratic Citizenship in an Era of Transnational Migration and Global Conflict." *Harvard Educational Review* 77 (3): 285–316.

Alba, Richard, and Victor Nee. 1997. "Rethinking Assimilation Theory for a New Era of Immigration. *International Migration Review* 31: 826–874.

Allen, Kieran. 1999. "Immigration and the Celtic Tiger: A Land of a Thousand Welcomes." In *The European Union and Migrant Labour,* ed. M. Cole and G. Dale. London: Berg.

Allport, Gordan. 1954. *The Nature of Prejudice.* Cambridge, MA: Addison-Wesley.

Altbach, Philip. 1978. "The Distribution of Knowledge in the Third World: A Case Study in Neo-Colonialism." In *Education and Colonialism* ed. Phillip Altbach and Gail Kelly. New Brunswick: Transaction Books.

Alvarez, Sonia E., Evelina Dagnino, and Arturo Escobar, eds. 1998. *Cultures of Politics/Politics of Cultures: Re-visioning Latin American Social Movements.* Boulder, CO: Westview.

Anderson, Gary L. 1998. "Toward Authentic Participation: Deconstructing the Discourses of Participatory Reforms in Education." *American Educational Research Journal* 35 (4): 571–603.

Ang, Ien. 1996. "The Curse of the Smile: Ambivalence and the 'Asian' Woman in Australian Multiculturalism." *Feminist Review* 52: 36–49.

Appadurai, Arjun. 1996. *Modernity at Large: Cultural Dimensions of Globalization.* Minneapolis: University of Minnesota Press.

———. 2000. "Grassroots Globalization and the Research Imagination." *Public Culture* 12 (1): 1–19.

Apple, Michael. 2000. *Official Knowledge: Democratic Education in a Conservative Age.* New York: Routledge.
————. 2001. *Educating the "Right Way": Markets, Standards, God and Inequality.* New York: Routledge.
————. 2005. "Are Markets in Education Democratic? Neoliberal Globalism, Vouchers and the Politics of Choice." In *Globalizing Education,* ed. M. Singh, J. Kenway, and M. Apple. New York: Peter Lang.
Apple, Michael. W., and James A. Beane, eds. 1995. *Democratic Schools.* Alexandria, VA: Association for Supervision and Curriculum Development.
Asimeng-Boahene, Lewis. 2003. "Understanding and Preventing Burnout among Social Studies Teachers in Africa. *Social Studies* 94 (2): 58–62.
Azevedo, José Clovis de. 2000. *Escola Cidadã: Desafios, Diálogos e Travessias.* Petrópolis: *Editora Vozes.*
Bahl, Ankur, Anna Johnson, and Carrie Seim. 2003. "Immigrants Face Climate of Fear." *The Chicago Reporter,* June. http://www.chicagoreporter.com/index.php/c/Cover_Stories/d/Immigrants_Face_Climate_of_Fear
————. 1995. "Intellectuals or Technicians? The Urgent Role of Theory in Educational Studies. *British Journal of Educational Studies* 43 (3): 255–271.
————. 2006. *Education Policy and Social Class: The Selected Works of Stephen J. Ball.* London: Routledge.
Bartlett, Lesley. 2005. "Dialogue, Knowledge, and Teacher-Student Relations: Freirean Pedagogy in Theory and Practice." *Comparative Education Review* 49 (3): 344–364.
————. 2007. "Human Capital or Human Connections? The Cultural Meanings of Education in Brazil." *Teachers College Record* 109 (7): 1613–1636.
————. forthcoming. *The Word and the World: The Cultural Politics of Literacy in Brazil.* Creskill, NJ: Hampton Press.
Bartlett, Lesley, Marla Frederick, Thaddeus Guldbrandsen, and Enrique Murillo, Jr. 2002. "The Marketization of Education: Public Schools for Private Ends." *Anthropology and Education Quarterly* 33 (1): 5–29.
Bennell, Paul. 2004. *Teacher Motivation and Incentives in Sub-Saharan Africa and Asia.* http://www.eldis.org/fulltext/dfidtea.pdf
Bereday, George. 1964. *Comparative Method in Education.* New York: Holt, Rinehart and Winston.
Berger, Peter, and Thomas Luckmann. 1967. *The Social Construction of Reality: A Treatise in the Sociology of Knowledge.* London: Penguin.
Berliner, D. C. 2005. *Carry It On: Fighting for Progressive Education in Neo-liberal Times.* Retrieved on January 30, 2008 from www.timeoutfortesting.org.
Bethke, Lynne, and Scott Braunschweig. 2003. *Global Survey on Education in Emergencies: Angola Country Report.* New York: Women's Commission for Refugee Women and Children.
Bhatt, Rakesh M. 2005. "Expert Discourses, Local Practices, and Hybridity: The Case of Indian Englishes." In *Reclaiming the Local in Language Policy and Practice,* ed. A. S. Canagarajah. Mahwah, NJ: Lawrence Erlbaum.
Blakemore, Priscilla. 1970. "Assimilation and Association in French Educational Policy and Practice: Senegal 1903–1939." In *Essays in the History of African*

Education, ed. M. Vincent, V. M. Battle, and C. H. Lyons. New York: Teachers College.

Blommaert, Jan. 2006. Language Policy and National Identity. In *An Introduction to Language Policy: Theory and Method*, ed. T. Ricento. Malden, MA & Oxford, UK: Blackwell.

Blommaert, Jan, and Jef Verschueren. 1998. *Debating Diversity: Analyzing the Discourse of Tolerance*. London; New York: Routledge.

Bonnett, Alastair. *1993. Radicalism, Anti-Racism and Representation*. London: Routledge.

Bouche, Denise. 1975. *"L'Enseignement dans les Territories Français de l'Afrique Ocidentale de 1817 a 1920"* [Schooling in French West Africa from 1817 to 1920]. PhD diss., University of Lille.

Bourdieu, Pierre. 1986. "The Forms of Capital." In *Handbook of Theory and Research for the Sociology of Education*, ed. J. Richardson. New York: Greenwood.

———. 2001. *Masculine Domination*. Cambridge, UK: Polity Press.

Bourdieu, Pierre, and Jean Claude Passeron. 1977. *Reproduction in Education, Society and Culture*. London: Sage.

Bray, Mark, and R. Murray Thomas. 1995. "Levels of Comparison in Educational Studies: Different Insights from Different Literatures and the Value of Multilevel Analyses." *Harvard Educational Review* 65 (3): 472–490.

Broadfoot, Patricia. 1999. "Stones from Other Hills May Serve to Polish the Jade of This One: Towards a Neo-Comparative 'Learnology' of Education." *Compare* 29 (3): 217–232.

Brock-Utne, Birgit. 2003. "Formulating Higher Education Policies in Africa: The Pressure from External Forces and the Neoliberal Agenda." *Journal of Higher Education in Africa* 1 (1): 24–56.

Bryan, Audrey. 2008. "The Co-articulation of National Identity and Interculturalism in the Irish Curriculum: Educating for Democratic Citizenship?" *London Review of Education* 6 (1): 47–58.

Buchen, Charlotte. 2008. "Pakistani American Stand up: Homeland Concerns Fuel Political Participation." *Frontline World*, October 28. http://www.pbs. org/frontlineworld/election2008/2008/10/pakistani-americans-stand.html

Buor, Sei Goryor. 2001. "Ethnonational Tensions in Liberian Education, 1944–1996." PhD diss., Loyola University.

Burawoy, Michael, ed. 2000. *Global Ethnography: Forces, Connections, and Imaginations in a Postmodern World*. Berkeley: University of California Press.

Burbules, Nicholas, and Carlos Alberto Torres, eds. 2000. *Globalization and Education: Critical Perspectives*. New York: Routledge.

Campbell, Mary L., and Frances M. Gregor. 2002. *Mapping Social Relations: A Primer in Doing Institutional Ethnography*. Aurora, ON: Garamond Press.

Campbell, Marie, and Ann Manicom. 1995. "Introduction." In *Knowledge, Experience, and Ruling Relations: Studies in the Social Organization of Knowledge*, ed. M. Campbell and A. Manicom. Toronto: University of Toronto Press.

Canagarajah, Attelstan S., ed. 2005. *Introduction to Reclaiming the Local in Language Policy and Practice*. Mahwah, NJ: Lawrence Erlbaum.

Carnoy, Martin. 2006. "Rethinking the Comparative and the International." *Comparative Education Review* 50 (4): 551–570.

Carnoy, Martin, and Diana Rhoten. 2002. "What Does Globalization Mean for Educational Change? A Comparative Approach." *Comparative Education Review* 46 (1): 1–7.

Carvalho, Jose. M. 2002. *Cidadania no Brasil: O Longo Caminho.* Rio de Janeiro: Civilização Brasileira.

Central Statistics Office. 2007. *Census 2006: Principal Demographic Results.* Dublin: Stationary Office.

Centre de Recherche et de Développement Pédagogiques (CRDP). 2007. *Al nashra al ih?sa?iyya. [Statistics bulletin].* Beirut: Ministry of Education CRDP. Retrieved January 15, 2008, from http://www.crdp.org/crdp/Arabic/ar-statistics/a_statisticpublication.asp. [In Arabic]

Cerrón-Palomino, Rodolfo. 1989. "Language Policy in Peru: A Historical Overview." *International Journal of the Sociology of Language* 77: 11–33.

Chafer, Tony. 1997. "Students and Nationalism: The Role of Students in the Nationalist Movement in Afrique Occidental Française (AOF), 1946–60." In *AOF: Réalitiés et Heritages: Sociétés Ouest-Africaines et Ordre Colonial, 1895–1960* [AOF Realities and Heritage: West African Societies and the Colonial Order], ed. C. Becker, S. Mbaye, and I. Thiuob. Dakar: Direction des Archives du Sénégal.

Chapman, David. 2002. *Corruption and the Education Sector.* Washington, DC: USAID.

Chauvet, Lisa, and Paul Collier. 2007. "Education in Fragile States." Paper commissioned for the *EFA Global Monitoring Report 2008: Education by 2015: Will We Make It?* Geneva: UNESCO.

Cleaver, Frances. 1999. "Paradoxes of Participation: Questioning Participatory Approaches to Development." *Journal of International Development* 11: 597–612.

Clifford, James, and George E. Marcus, eds. 1986. *Writing Culture: The Poetics and Politics of Ethnography.* Berkeley: University of California Press.

Clinchy, Evans, ed. 2000. *Creating New Schools: How Small Schools Are Changing American Education.* New York: Teachers College.

Conflict and Education Research Group. 2008. *Education and Fragility.* Desk study commissioned by the Inter-Agency Network for Education in Emergencies (INEE). http://ineesite.org/uploads/documents/store/doc_1_FINAL-Desk_Study_Education_and_Fragility_CERG2008.pdf

Connolly, Paul. 1998. *Racism, Gender Identities and Young Children: Social Relations in a Multi-Ethnic, Inner City Primary School.* London: Routledge.

———. 2000. "What Now for the Contact Hypothesis? Towards a New Research Agenda." *Race, Ethnicity and Education* 3 (2): 169–193.

———. 2006. " 'It Goes Without Saying (Well, Sometimes).' Racism, Whiteness and Identity in Northern Ireland." In *The New Countryside? Ethnicity, Nation and Exclusion in Contemporary Rural Britain,* ed. J. Agyeman and S. Neal. Bristol: Policy Press.

Comisión de la Verdad y la Reconciliación. (2003). *Informe final.* http://www. cverdad.org.pe/ingles/ifinal/index.php

Cooper, Frederick, and Randall Packard, eds. 1997. *International Development and the Social Sciences: Essays on the History and Politics of Knowledge.* Berkeley: University of California Press.

Coquery-Vidrovitch, Catherine. 1985. "The Colonial Economy of the Former French, Belgian and Portuguese Zones 1914–35." In *General History of Africa: Africa under Colonial Domination 1880–1935,* ed. A. A. Boahen. Paris: UNESCO and Heinemann.

Coulter, Colin. 2003. "The End of Irish History? Introduction." In *The End of Irish History?: Critical Approaches to the Celtic Tiger,* ed. C. Coulter and S. Coleman. Manchester: Manchester University Press.

Crisis States Workshop. 2006. *Crisis, Fragile and Failed States: Definitions Used by the CSRC.* http://www.crisisstates.com/download/drc/FailedState.pdf

Crossley, Michael, and Graham Vulliamy. 1984. "Case Study Research Methods and Comparative Education." *Comparative Education* 20 (2): 193–207.

Dabbagh, Salah, George Deeb, Farid el Khazen, and Maroun M. Kisirwani, with a Preface by Bahige Tabbarah. 1997. "The Lebanese Constitution, 1990." *Arab Law Quarterly* 12 (2): 224–261.

Dale, Roger. 2000. "Globalization: A New World for Comparative Education?" In *Discourse Formation in Comparative Education,* ed. J. Schriewer. Oxford: Peter Lang.

———. 2003. "Positive Post-Structuralism? A Response to Ninnes and Burnett." *Comparative Education* 39 (3): 307–309.

Davies, Lynn. 2004. *Education and Conflict: Complexity and Chaos.* London: Falmer Press.

de Certeau, Michel. 1984. *The Practice of Everyday Life.* Berkeley: University of California Press.

De la Cadena, Marisol. 2000. *Indigenous Mestizos: The Politics of Race and Culture in Cuzco, Peru, 1919–1991.* Durham, NC: Duke University Press.

Department of Education and Science [DES] 2002. *Promoting Anti-Racism and Interculturalism in Education: Draft Recommendations towards a National Action Plan.* Dublin: DES.

Department of Justice, Equality and Law Reform. 2005. *Planning for Diversity: The National Action Plan against Racism 2005–2008.* Dublin: Department of Justice Equality and Law Reform.

Deutsch, Morton. 2006. "Justice and Conflict." In *The Handbook of Conflict Resolution: Theory and Practice.* 2nd edition, ed. M. Deutsch and P. Coleman. San Francisco: Jossey-Bass.

Deutsche Gesellschaft für Technische Zusammenarbeit. 2008. *Educational Reform Teacher Training in Intercultural Bilingual Education and Intercultural Bilingual Education in Five Andes Nations.* http://www.gtz.de/en/weltweit/ lateinamerika-karibik/14054.htm

Devereux, Eoin, and Michael Breen. 2004. "No Racists Here: Public Opinion and Media Treatment of Asylum Seekers and Refugees." In *Political Iissues in Ireland Today,* ed. N. Collins and T. Cradden. Manchester: Manchester University Press.

Devereux, Eoin, Michael Breen, and Amanda Haynes. 2005. " 'Smuggling Zebras for Lunch': Media Framing of Asylum Seekers in the Irish Print Media." *Etudes Irlandaises* 30 (1): 109–130.

———. 2006a. "Citizens, Loopholes and 'Maternity Tourists': Media Framing of the 2004 Citizenship Referendum." In *Uncertain Ireland,* ed. M. Corcoran and M. Peillon. Dublin: IPA.

———. 2006b. *Media and Migration.* University of Antwerp: Dialogue Series No. 6.

Devine, Dympna. 2005. "Welcome to the Celtic Tiger? Teacher Responses to Immigration and Ethnic Diversity in Irish Schools." *International Studies in the Sociology of Education* 15 (1): 49–70.

Dyer, Caroline, and Archana Choksi, with Vinita Awasty, Uma Iyer, Renu Moyade, Neerja Nigam, and Neetu Purohit. 2002. "Democratising Teacher Education Research in India." *Comparative Education* 38: 337–351.

Eastwood, Lauren E. 2006. "Making the Institution Ethnographically Accessible: UN Document Production and the Transformation of Experience." In *Institutional Ethnography as Practice,* ed. D. E. Smith. Lanham, MD: Rowman and Littlefield.

Escobar, Arturo. 1995. *Encountering Development: The Making and Unmaking of the Third World.* Princeton, NJ: Princeton University Press.

———. 2000. "Place Power and Networks on Globalization and Post Development." In *Redeveloping Communication for Social Change: Theory, Practice and Power,* ed. K. G. Wilkins. Lanham, MD: Rowman and Littlefield.

Esteva, Gustavo, and Madhu Suri Prakash. 1998. *Grassroots Post-Modernism: Remaking the Soil of Cultures.* London and New York: Zed Books.

EUROSTAT. 2006. *Euro-zone Unemployment Stable at 8.8%: EU15 Steady at 8.0%.* http://europa.eu.int/comm/eurostat/Public/datashop/print-product/EN?catal ogue=Eurostatandproduct=3-02032004-EN-AP-ENandmode=download

Fairclough, Norman. 1995. *Critical Discourse Analysis.* London: Longman.

———. 2003. *Analyzing Discourse: Textual Analysis for Social Research.* New York: Routledge.

Ferguson, James. 1994. *The Anti-Politics Machine: "Development," Depoliticization, and Bureaucratic Power in Lesotho.* Minneapolis, MN: University of Minnesota Press.

Fine, Michelle. 1994. *Chartering Urban School Reform: Reflections on Public High Schools in the Midst of Change.* New York: Teachers College.

———. 2000. "A Small Price to Pay for Justice." In *A Simple Justice,* ed. W. Ayers. New York: Teachers College.

———. 2005. "Not in Our Name." *Rethinking Schools* 19 (4): 1–6.

Fine, Michelle, and Linda Powell. 2001. "Small Schools: An Anti-Racist Intervention in Urban America." In *Racial Profiling and Punishment in U.S. Public Schools,* ed. T. Johnson, J. Boyden, and W. Pitz. Oakland, CA: Applied Research Center.

Fine, Michelle, and Janis I. Somerville, eds. 1998. *Small Schools: Big Imaginations.* Chicago: Cross City Campaign for Urban Education Reform.

Foley, Douglas. 1977. "Anthropological Studies of Schooling in Developing Countries: Some Recent Findings and Trends." *Comparative Education Review* 21 (2–3): 311–328.

Frayha, Nemer. 2004. "Developing Curriculum as a Means to Bridging National Divisions in Lebanon." In *Education, Conflict, and Social Cohesion*, ed. S. Tawil and A. Harley. Geneva: UNESCO International Bureau of Education.

Freeland, Jane. 1995. "'Why Go to School to Learn Miskitu?': Changing Constructs of Bilingualism, Education and Literacy among the Miskitu of Nicaragua's Atlantic Coast." *International Journal of Educational Development* 15 (3): 245–261.

———. 1996. "The Global, the National, and the Local: Forces in the Development of Education for Indigenous Peoples—the Case of Peru." *Compare* 26 (2): 167–195.

Freire, Paulo. 1994. *Pedagogy of the Oppressed*. New York: Continuum.

French, Brigittine. 1999. "Imagining the Nation: Language Ideology and Collective Identity in Contemporary Guatemala." *Language and Communication* 19: 277–287.

Fuglesang, Minou. 2002. "Voices on Femina-HIP Magazine—Using Edutainment to Promote Open Discussion about Sexuality and Risk Behavior." In *One Step Further—Responses to HIV/AIDS, Side Studies* 7, ed. A. Sisask. Stockholm, Sweden: SIDA.

Gandin, Luis A. 2002. "Democratizing Access, Governance, and Knowledge: The Struggle for Educational Alternatives in Porto Alegre, Brazil." PhD diss., University of Wisconsin, Madison.

Gandin, Luis. A., and Michael Apple. 2002. "Thin versus Thick Democracy in Education: Porto Alegre and the Creation of Alternatives to Neo-Liberalism." *International Studies in Sociology of Education* 12 (2): 99–115.

García, Ofelia., and Zeena Zakharia. forthcoming. "Language, Ethnic Identities, and the Education of Language Minority Children." In *Handbook of Research on Ethnic Identity and Development*, ed. R. Hernández Sheets and P. R. Portes. Mahwah, NJ: Lawrence Erlbaum.

Garner, Steve. 2004. *Racism in the Irish Experience*. London: Pluto.

Geertz, Clifford. 1993 (1973). *The Interpretation of Cultures: Selected Essays*. London: Fontana.

Genro, Tarso. 1997. *Orçamento Participativo: A Experiência de Porto Alegre*. São Paulo: Editora Fundação Perseu Abramo.

Gentili, Pablo. 2000. "A Complexidade do óbvio: A Prvatização e seus Significados no Campo Educacional." In *A Escola Cidadã no Contexto da Globalização*, 4th edition, ed. L. Heron da Silva. Petrópolis: Editora Vozes.

Gillborn, David. 1995. *Racism and Antiracism in Real Schools: Theory, Policy, Practice*. Buckingham, England and Philadelphia: Open University Press.

Gingrich, Andre. 2002. *Anthropology, by Comparison*. New York: Routledge.

Gootman, Elissa. 2006a. "36 More Small Schools Due in September, Mayor Says." *New York Times*, February 2. www.nytimes.org

———. 2006b. "Worst Graduation Rates Are in New York, Study Says." *New York Times*, February 16. www.timeoutfromtesting.org

Green, Maia. 2003. "Globalizing Development in Tanzania: Policy Franchising through Participatory Project Management." *Critique of Anthropology* 23 (2): 123–143.

Greene, Maxine. 2005. *Teaching in a Moment of Crisis: The Space of Imagination.* New York: Teachers College.

Grilo, Luisa. 2006. *"A Educação em Angola: Desafios da Reconstrução* [Education in Angola: Reconstruction challenges]." *Educar sem Fronteiras* [*Education without borders*] 3. Viana do Castelo, Portugal: Escola Superior de Educação de Viana do Castelo.

Habte, Aklilu, and Teshome Wagaw. 1993. "Education and Social Change." In *General History of Africa: Africa Since 1935,* ed. A. Mazrui. Paris: UNESCO and Heinemann.

Hage, Ghassan. 1998. *White Nation: Fantasies of White Supremacy in a Multicultural Society.* Annandale, New South Wales and UK: Pluto.

———. 2003. *Against Paranoid Nationalism: Searching for Hope in a Shrinking Society.* Annandale, New South Wales and UK: Pluto.

———. 2004. "The Ethnography of Imagined Communities: The Cultural Production of Sikh Ethnicity in Britain." *Annals of the American Academy of Political and Social Science* 595: 108–121.

Hall, Stuart. 1996. Introduction: "Who Needs Identity?" In *Questions of Cultural Identity,* ed. S. Hall and P. D. Gay. London: Sage.

Hans, Nicholas. 1949. *Comparative Education: A Study of Educational Factors and Traditions.* London: Routledge and Kegan Paul.

Hantzopoulos, Maria. 2004. "The Impact of Standardized Testing on the Deliverance of Post-Conflict Education: A Case Study of One New York City High School." *Education in Emergencies: A Columbia University Graduate Student Publication.* New York: Columbia University.

———. 2008. "Sizing Up Small: An Ethnographic Case Study of a Critical Small High School in New York City." PhD diss., Teachers College, Columbia University.

Harber, Clive. 1996. *Small Schools and Democratic Practice.* Nottingham: Education Heuristics Press.

Harvey, David. 1989. *The Condition of Postmodernity: An Enquiry into the Origins of Cultural Change.* Cambridge, MA: Basil Blackwell.

Hayhoe, Ruth, and Karen Mundy. 2008. "Introduction to Comparative and International Education: Why Study Comparative Education?" In *Comparative and International Education: Issues for Teachers,* ed. K. Mundy, K. Bickmore, R. Hayhoe, M. Madden, and K. Madjidi. Toronto: Canadian Scholars' Press; New York and London: Teachers College.

Henriques, J. 1984. *Changing the Subject: Psychology, Social Regulation and Subjectivity.* London; New York: Methuen.

Heyman, Richard. 1979. "Comparative Education from an Ethnomethodological Perspective." *Comparative Education* 15 (3): 241–249.

Hickey, Samuel, and Giles Mohan. 2005. "Relocating Participation Within a Radical Politics of Development." *Development and Change* 36 (2): 237–262.

Holt, Mike. 1996. "Divided Loyalties: Language and Ethnic Identity in the Arab World." In *Language and Identity in the Middle East and North Africa,* ed. Y. Suleiman. Richmond, Surrey: Curzon Press.

Hornberger, Nancy. H. 2000. "Bilingual Education Policy and Practice in the Andes: Ideological Paradox and Intercultural Possibility." *Anthropology and Education Quarterly* 31 (2): 173–201.

International Crisis Group. 2003. *Dealing with Savimbi's Ghost: The Security and Humanitarian Challenges in Angola.* Luanda/Brussels: International Crisis Group.

International Monetary Fund. [IMF]. 2003. *Angola: 2003 Article IV Consultation* (Country Report No. 03/291) http://www.imf.org/external/pubs/cat/longres. cfm?sk=16871.0

Johannessen, Eva. 2000. *Evaluation of Teacher Emergency Package (TEP) in Angola.* Oslo: Educare.

Joint Committee on Education and Science. 2004. *Educational Supports: Discussion with Department of Education and Science and NCCA.* http://debates.oireachtas. ie/DDebate.aspx?F=EDJ20070222.XML&Ex=All&Page=2

Joseph, John. E. 2004. *Language and Identity: National, Ethnic, Religious.* Basingstoke, Hampshire, and New York: Palgrave Macmillan.

Kaiser Family Foundation and UNAIDS. 2004. *The Media and HIV/AIDS: Making a Difference.* http://www.kff.org/hivaids/loader.cfm?url=/commonspot/ security/getfile.cfm&PageID=29879

Kandel, Isaac L. 1933. *Comparative Education.* Boston: Houghton Mifflin.

Kapoor, Ilan. 2002. "The Devil's in the Theory: A Critical Assessment of Robert Chambers' Work on Participatory Development." *Third World Quarterly* 23 (1): 101–117.

Kenway, Jane. 1993. "Marketing Education in the Post-Modern Age." *Journal of Educational Policy* 8(2): 105–122.

Kim, Jennifer, Naseem Kourosh, Jayne Huckerby, Kobi Leins, and Smita Narula, S. 2007. "Americans on Hold: Profiling, Citizenship, and the 'War on Terror.'" *Center for Human Rights and Global Justice,* NYU School of Law [Online]. Available at http://www.chrgj.org/docs/AOH/AmericansonHoldReport.pdf

King, Kendall A. 2001. *Language Revitalization Processes and Prospects: Quechua in the Ecuadorian Andes.* Clevedon, UK: Multilingual Matters.

Ki-Zerbo, Joseph. 1974. "Education and Development in the Bellagio Conference Papers." *Education and Development Reconsidered.* New York: Praeger.

Klonsky, Michael, and Susan Klonsky. 2008. *Small Schools: Public School Reform Meets the Ownership Society.* New York: Routledge.

Know Racism. 2005. "The National Anti-Racism Awareness Programme, Final Report on Activities 2001–2003." Unpublished document.

LaFraniere, Sharon. 2007. October 14. "As Angola Rebuilds, Most Find Their Poverty Persists." *New York Times,* October 14. http://www.nytimes. com/2007/10/14/world/africa/14angola.html?_r=1g

Latour, Bruno. 1987. *Science in Action: How to Follow Scientists and Engineers through Society.* Cambridge, MA: Harvard University Press.

———. 1995. "Social Theory and the Study of Computerized Work Sites." In *Information Technology and Changes in Organizational Work,* ed. N. J. Orlinokowski and G. Walsham. London: Chapman and Hall.

———. 2005. *Reassembling the Social: An Introduction to Actor-Network Theory.* Oxford: Oxford University Press.

Lemke, Jay L. 1995. *Textual Politics: Discourse and Social Dynamics.* London: Taylor and Francis.

Levett, Ann, Amanda Kottler, Erica Burman, and Ian Parker. 1997. *Culture, Power & Difference: Discourse Analysis in South Africa.* Cape Town: University of Cape Town Press.

Levi-Strauss, Claude. 1966. *The Savage Mind.* Translated by J. Weightman and D. Weightman. Chicago: University of Chicago Press.

Levin, Henry M. 2006. "Commentary on Carnoy." *Comparative Education Review* 50 (4): 571–580.

Levinson, Bradley A. 1999. "Resituating the Place of Educational Discourse in Anthropology." *American Anthropologist* 101 (3): 594–604.

Levinson, Bradley A. U., and Margaret Sutton. 2001. "Introduction: Policy as/ in Practice: A Sociocultural Approach to the Study of Educational Policy." In *Policy as Practice: Toward a Comparative Sociocultural Analysis of Educational Policy*, ed. M. Sutton and B. A. U. Levinson. Westport, CT: Ablex.

Levister, Chris. 2005. NCLB Fails Blacks, Latinos in Urban Schools: Ds and Fs for NCLB. *The Black Voice Online*, www.timeoutfromtesting.org

Levitt, Jeremy. 2005. *The Evolution of Deadly Conflict in Liberia: From Paternalism to State Collapse.* Durham, NC: Carolina Academic Press.

Ley General de Educación. 2003. *Principios de la Educación.* http://www.minedu.gob.pe/normatividad/leyes/ley_general_de_educacion2003.doc

Lieb, Stephen. 1991. *Principles of Adult Learning.* http://honolulu.hawaii.edu/intranet/committees/FacDevCom/guidebk/teachtip/adults-2.htm

Linz, Juan, and Alfred Stepan. 1996. Toward Consolidated Democracies. *Journal of Democracy* 7 (2): 14–33.

Little, Judith Warren. 1982. "Norms of Collegiality and Experimentation: Workplace Conditions of School Success." *American Educational Research Journal* 19 (3): 325–340.

Lodge, Anne, and Kathleen Lynch, eds. 2004. *Diversity at School.* Dublin: IPA/ Equality Authority.

López, Luis Enrique. 1997. "*La Eficacia y la Validez de lo Obvio: Lecciones Aprendidas desde la Evaluación de los Procesos Educativos Bilingües.*" In *Multilingüismo y Educación Bilingüe en América y España*, ed. J. Calvo Pérez and J. C. Godenzzi. Cuzco: Centro de Estudios Regionales Andinos Bartolomé de las Casas.

Lopez, Nancy. 2003. *Hopeful Girls, Troubled Boys: Race and Gender Disparity in Urban Education.* New York: Routledge.

Loyal, Steve. 2003. "Welcome to the Celtic Tiger: Racism, Immigration and the State." In *The End of Irish History: Critical Reflections on the Celtic Tiger*, ed. C. Coulter and S. Coleman. Manchester: Manchester University Press.

Lunn, Joe. 1999. *Memoirs of the Malestrom: A Senegalese Oral History of the First World War.* Oxford: James Currey.

Maira, Sunaina Marr. 2002. *Desis in the House.* Philadelphia: Temple University Press.

———. 2004. "Imperial Feelings: Youth Culture, Citizenship, and Globalization." In *Globalization: Culture and Education in the New Millennium*, ed. M. M. Suárez-Orozco and D. B. Qin-Hilliard. Berkeley: University of California Press.

Makdisi, Ussama. 2000. *The Culture of Sectarianism: Community, History, and Violence in Nineteenth-Century Ottoman Lebanon.* Berkeley: University of California Press.

Makongo, Japhet, and Marjorie Mbilinyi. 2003. *The Challenge of Democratizing Education Governance at the Local Level.* HakiElimu Working Paper Series No. 2003.9. Dar es Salaam: HakiElimu.

Malaquias, Assis. 2007. *Rebels and Robbers: Violence in Post-Conflict Angola.* Stockholm, Sweden: Elanders Gotab AB.

Mannheim, Bruce. 1984. *"Una Nación Acorralada*: Southern Peruvian Quechua language Planning and Politics in Historical Perspective." *Language in Society* 13 (3): 291–309.

Marcus, George. 1989. "Imagining the Whole: Ethnography's Contemporary Efforts to Situate Itself." *Critique of Anthropology* 9 (3): 7–30.

———. 1998. *Ethnography through Thick and Thin.* Princeton, NJ: Princeton University Press.

Marcus, George. E., and Michael M. J. Fisher. 1986. *Anthropology as Cultural Critique: An Experimental Moment in the Social Sciences.* Chicago and London: University of Chicago Press.

Marginson, Simon, and Marcela Mollis. 2001. "'The Door Opens and the Tiger Leaps': Theories and Reflexivities of Comparative Education for a Global Millennium." *Comparative Education Review* 45 (4): 581–615.

Masemann, Vandra. 1976. "Anthropological Approaches to Comparative Education." *Comparative Education Review* 20 (3): 368–380.

———. 1982. "Critical Ethnography in the Study of Comparative Education." *Comparative Education Review* 26 (1): 1–15.

———. 1990. "Ways of Knowing: Implications for Comparative Education." *Comparative Education Review* 34 (4): 465–473.

Masinda, Mambi, and Muhindo Muhesi. 2004. "Children and Adolescents' Exposure to Traumatic War Stressors in the Democratic Republic of Congo." *Journal of Child and Adolescent Mental Health* 16: 25–30.

Massialas, Byron G., and Samir A. Jarrar. 1991. *Arab Education in Transition: A Source Book.* New York and London: Garland.

Mathis, William J. 2004. "The Federal No Child Left Behind Act: What Will It Cost States?" *Spectrum: The Journal of State Government* 77 (2): 8–10. http://www.csg.org/pubs/Documents/spec_sp04.pdf

Mazrui, Ali. 1975. "The African University as a Multinational Corporation: Problems of Penetration and Dependency." *Harvard Educational Review* 45 (2): 191–210.

McCarthy, Cameron. 1993. "After the Canon: Knowledge and Ideological Representation in the Multicultural Discourse on Curriculum Reform." In *Race, Identity and Representation in Education,* ed. C. McCarthy and W. Crichlow. New York and London: Routledge.

McCarthy, Cameron, Warren Crichlow, Greg Dimitriadis, and Nadine Dolby. 2005. "Introduction: Transforming Contexts, Transforming Identities: Race and Education in the New Millennium." In *Race, Identity, and Representation in*

Education, ed. C. McCarthy, W. Crichlow, G. Dimitriadis, and N. Dolby. New York: Routledge.

McCarty, Teresa. 2004. "Language Education Policies in the United States." In *Medium of Instruction Policies: Which Agenda? Whose Agenda?* ed. J. W. Tollefson and A. B. M. Tsui. Mahwah, NJ: Lawrence Erlbaum.

McDaid, Rory. 2007. "New Kids on the Block." In *Beyond Educational Disadvantage,* ed. P. Downes and A. L. Gilligan. Dublin: IPA.

McGinnity, Frances, Philip O'Connell, Emma Quinn, and James Williams. 2006. *Migrants' Experience of Racism and Discrimination in Ireland. Final Report to the European Monitoring Centre on Racism and Xenophobia.* Dublin: ESRI.

McVeigh, Robbie, and Ronit Lentin. 2002. "Situated Racisms: A Theoretical Introduction." In *Racism and Anti-Racism in Ireland,* ed. R. Lentin and R. McVeigh. Belfast: Beyond the Pale.

Medeiros, Emilie. 2007. "Integrating Mental Health into Post-Conflict Rehabilitation: The Case of Sierra Leonean and Liberian 'Child Soldiers.'" *Journal of Health Psychology* 12 (3): 498–504.

Meier, Deborah. 1995. *The Power of Their Ideas: Lessons for America from a Small School in Harlem.* Boston: Beacon.

———. 2000a. "Educating for a Democracy." In *Will Standards Save Public Education?* ed. J. Cohn and J. Rogers. Boston: Beacon.

———. 2000b. "The Crisis of Relationships." In *A Simple Justice,* ed. W. Ayers. New York: Teachers College.

———. 2002. *In Schools We Trust.* Boston: Beacon.

———. 2004. "NCLB and Democracy." In *Many Children Left Behind,* ed. D. Meier and G. Wood. Boston: Beacon.

Miller, Jeremy. 2008. "Tyranny of the Test: One Year as a Kaplan Coach in the Public Schools." *Harpers,* September, 35–46.

Ministerio de Educación. 2003. *Lineamientos de la Educación Bilingüe Intercultural.* Urubamba, Perú: Instituto Pedagógico La Salle.

———. 2008. *Unidad de Medición de la Calidad, UMC.* http://www2.minedu. gob.pe/umc/noticiacompleta_index.php?v_codigo=1

Ministry of Education Republic of Liberia. 2007. *Liberian Primary Education Recovery Program Prepared for the Fast Track Initiative.* http://www.poledakar. org/IMG/Liberia_EFA-FTI.pdf

Monkman, Karen, and Mark Baird. 2002. "Review: Educational Change in the Context of Globalization." *Comparative Education Review* 46 (4): 497–508.

Montgomery, Kenneth. 2005. *"A Better Place to Live": National Mythologies, Canadian History Textbooks, and the Reproduction of White Supremacy.* Ottawa: University of Ottawa.

Morrow, Raymond, and Carlos A. Torres. 2000. "The State, Globalization, and Educational Policy." In *Globalization and Education: Critical Perspectives,* ed. N. Burbules and C. Torres. New York: Routledge.

Mosse, David. 2005. *Cultivating Development: An Ethnography of Aid Policy and Practice.* London: Pluto.

National Center for Educational Research and Development. 1995. *New Framework for Education in Lebanon.* Beirut: NCERD.

National Commission on Excellence in Education. 1983. *A Nation at Risk: The Imperative for Educational Reform*. Washington, DC: U.S. Department of Education.

National Council for Curriculum and Assessment [NCCA]. 2005. *Intercultural Education in the Post-Primary School: Guidelines for Schools*. Dublin: NCCA.

Ndiaye, Falilou. 2000. *"La Condition des Universitaires Sénégalais"* [The Condition of Senegalese Universities]. In *The Dilemma of Post-Colonial Universities: Elite Formation and the Restructuring of Higher Education in sub-Saharan Africa*, ed. Y. Lebeau and M. Ogunsanya. Ibadan: French Institute for Research in Africa.

Ndiaye, Honore-Georges. 2003. "Senegal." In *African Higher Education: An International Reference Handbook*, ed. D. Teferra and P. Altbach. Bloomington: Indiana University Press.

Nieto, Sonia. 2000. "A Gesture towards Justice: Small Schools and the Promise of Equal Education." In *A Simple Justice*, ed. W. Ayers, M. Klonsky, and G. Lyon. New York: Teachers College.

Ninnes, Peter, and Gregory Burnett. 2003. "Comparative Education Research: Poststructuralist Possibilities." *Comparative Education* 39 (3): 279–297.

Ninnes, Peter, and Sonia Mehta. 2004. *Re-Imagining Comparative Education*. New York: Routledge.

Ní Shuinear, Sinead. 1994. "Irish Travellers, Ethnicity and the Origins Question." In *Irish Travellers: Culture and Ethnicity*, ed. M. McCann, S. O Siochain, and J. Ruane. Belfast: Queens University Press.

No Child Left Behind. 2004. http://www.ed.gov/nclb

Noah, Harold, and Max Eckstein. 1969. *Toward a Science of Comparative Education*. New York: Macmillan.

———. 1998. *Doing Comparative Education: Three Decades of Collaboration*. Hong Kong: University of Hong Kong Press.

Noam, Gil G., Beth M. Miller, and Susanna Barry. 2002. "Youth Development and After-School Time: Policy and Programming in Large Cities." In *Youth Development and After-School Time: A Tale of Many Cities. New Directions for Youth Development, Number 94*, ed. G. G. Noam and B. M. Miller. San Francisco: Jossey-Bass.

Nonini, Donald, and Aihwa Ong. 1997. "Introduction: Chinese Transnationalism as an Alternative Modernity." In *Ungrounded Empires: The Cultural Politics of Modern Chinese Transnationalism*, ed. A. Ong and D. Nonini. New York: Routledge.

Norwegian Refugee Council. 1998. *Brief History and Main Features of the Teacher Emergency Package (TEP) for Angola: 1995–1998*. Luanda, Angola: Norwegian Refugee Council.

———. 2002. *Country Strategy: Angola 2002–2004*. Luanda, Angola: Norwegian Refugee Council.

Nowlan, Emer. 2008. "Underneath the Band-Aid: Supporting Bilingual Students in Irish Schools." *Irish Educational Studies* 27 (3): 253–266.

O'Donnell, Guillermo A. 1996. "Illusions about Consolidation." *Journal of Democracy* 7 (2): 34–51.

Office of the Minister for Integration [OMI]. 2008. *Migration Nation: Statement on Integration Strategy and Diversity Management*. Dublin: OMI.

"Parents, kids getting shut out of free help: Tutoring companies and advocates point to unkempt promise." 2006. *CNN.com*. Retrieved February 17, 2006 from http://www.cnn.com/2006/EDUCATION/02/17/school.tutors.ap/index. html

Parker, David. 2001. "The Chinese Takeaway and the Diasporic Habitus: Space, Time and Power Geometries." In *Unsettled Multiculturalisms: Diasporas, Entanglement, Transruptions*, ed. B. Hesse. New York: Palgrave Macmillan.

Parpart, Jane L. 1995. "Deconstructing the Development 'Expert.'" In *Feminism, Postmodernism, Development*, ed. M. H. Marchand and J. L. Parpart. London and New York: Routledge.

Piot, Charles. 1999. *Remotely Global: Village Modernization in West Africa*. Chicago and London: University of Chicago Press.

Pont, Raul. 2000. "*Democracia Representativa e Democracia Participativa*. In *Por uma Nova Esfera Pública: A Experiência do Orçamento Participativo*," ed. N. B. Fischer and J. Moll. Petrópolis: *Editora Vozes*.

Popkewitz, Thomas S., and Miguel A. Pereyra. 1993. "Reform Practices in Teacher Education." In *Changing Patterns of Power: Social Regulation and Teacher Education Reform*, ed. T. S. Popkewitz. Albany: State University of New York Press.

Powell, Linda. 2002. *Small Schools and the Issue of Race*. New York: Bank Street College of Education Occasional Paper Series.

Powell, Michael. 2003. "An Exodus Grows in Brooklyn: 9/11 Still Rippling through Pakistani Neighborhood" *Washington Post*, May 29, p. A01.

Psacharopolous, George. 1980. *Higher Education in Developing Countries: A Cost-Benefit Analysis*. World Bank Staff Working Paper No. 440, Washington, DC: World Bank.

Purkayastha, Bandana. 2005. *Negotiating Ethnicity: Second-Generation South Asian Americans Traverse a Transnational World*. New Jersey: Rutgers University Press.

Raduntz, Helen. 2005. "The Marketization of Education within the Global Capitalist Economy." In *Globalizing Education*, ed. M. Singh, J. Kenway, and M. Apple. New York: Peter Lang.

Rahnema, Majid. 1992. "Poverty." In *The Development Dictionary: A Guide to Knowledge and Power*, ed. Wolfgang Sachs. London: Zed Books.

Rawls, John. 1971. *A Theory of Justice*. Cambridge, MA: Harvard University Press.

Raywid, Mary Anne. 1993. "Finding Time for Collaboration." *Educational Leadership* 51 (1): 30–34.

Reddy, Shravanti. 2002. *Watchdog Organization Struggles to Decrease UN Bureaucracy*. http://www.globalpolicy.org/ngos/ngo-un/rest-un/2002/1029watchdog.htm

Reinhold, Susan. 1994. "Local Conflict and Ideological Struggle: 'Positive Images' and 'Section 28.'" PhD diss., University of Sussex.

ReliefWeb. 2004. *Reference Map of Angola*. http://www.reliefweb.int/rw/rwb.nsf/ db900SID/SKAR-64GDNK?OpenDocument

República de Angola. 2005. *Programa de Relançamento da Alfabetização e Recuperação do Atraso Escolar, 2006–2015* [Literacy and accelerated learning program launch, 2006–2015]. Luanda, Angola: *Ministério da Educação*.

Rizvi, Fazal. 1991. "The Idea of Ethnicity and the Politics of Multicultural Education." In *Power and Politics in Education*, ed. D. Dawkins. London: Falmer Press.

Robinson-Pant, Anna. 2001. "Development as Discourse: What Relevance to Education?" *Compare* 31 (3): 311–328.

Rodriguez, Louie, and Gilbeo Q. Conchas. 2008. *Small Schools and Urban Youth: Using the Power of School Culture to Engage Students.* Thousand Oaks, CA: Sage.

Roman, Leslie G., and Timothy Stanley. 1997. "Empires, Emigres and Aliens: Young People's Negotiations of Official and Popular Racism in Canada." In *Dangerous Territories: Struggles for Difference and Equality in Education,* ed. L. Roman and L. Eyre. New York: Routledge.

Rose, Pauline, and Martin Greeley. 2006. "Education in Fragile States: Capturing Lessons and Identifying Good Practice." Unpublished report prepared for the DAC Fragile States Group. http://www.ineesite.org/core_references/ Education_in_ Fragile_States.pdf

Sabatier, Peggy R. 1978. "Elite Education in French West Africa: The Era of Limits, 1903–1945." *International Journal of African Historical Studies* 11 (2): 247–266.

Sadler, Michael. 1900. "How Far Can We Learn Anything of Practical Value from the Study of Foreign Systems of Education?" In *Selections from Michael Sadler,* ed. J. H. Higginson. Liverpool: Dejall and Meyorre.

Samoff, Joel. 1999. "Education Sector Analysis in Africa: Limited National Control and Even Less National Ownership." *International Journal of Educational Development* 19 (4–5): 249–272.

Samoff, Joel, and Bidemi Carroll. 2004. "The Promise of Partnership and Continuities of Dependence: External Support to Higher Education in Africa." *African Studies Review* 47 (1): 67–199.

Samuels, Michael A. 1970. *Education in Angola, 1878–1914: A History of Culture Transfer and Administration.* New York: Teachers College.

Sarroub, Loukia K. 2005. *All American Yemeni Girls: Being Muslim in a Public School.* Philadelphia: University of Pennsylvania Press.

Saulny, Susan. 2004. "City High School Students Lag in Regents Test Scores." *New York Times.* www.timeoutfromtesting.org

Save the Children. 2006. *Rewrite the Future: Education for Children in Conflict-Affected Countries.* http://www.savethechildren.org/publications/reports/ RewritetheFuture-PolicyReport-1.pdf

Seck, A. 1994. *Concertation nationale sur l'enseignement supérieur au Sénégal: Rapport Definitif.* [Final CNES Report]. Dakar: CNES.

Secretária Municipal de Educação Prefeitura Municipal de Porto Alegre. 2006. *Eleição de Conselhos Escolares.* http://www2.portoalegre.rs.gov.br/smed/default. php?p_secao=98

Sen, Amartya. 2006. "What Clash of Civilizations? Why Religious Identity Isn't Destiny." *Slate,* March 29. http://www.slate.com/id/2138731/

Senge, Peter M. 1991. *The Fifth Discipline: The Art and Practice of the Learning Organization.* London: Century Business.

Shaaban, Kassim, and Ghazi Ghaith. 1999. "Lebanon's Language-in-Education Policies: From Bilingualism to Trilingualism." *Language Problems and Language Planning* 23 (1): 1–16.

Shehab, Soheil. 2006. "350 Schools Completely Destroyed or Damaged because of the Aggression: Lebanon Commends the Support of the UAE as It Confronts the Challenges of the New Academic Year." *Al Bayan*, October 16. http://www.albayan.ae/servlet/Satellite?c=Article&cid=1158495929288&pagename=AlbaYan%2FArticle%2FFullDetail

Shore, Cris, and Susan Wright, eds. 1997. *Anthropology of Policy: Critical Perspectives on Governance and Power*. New York: Routledge.

Sinclair, Margaret. 2001. "Education in Emergencies." In *Learning for a Future: Refugee Education in Developing Countries*, ed. J. Crisp, C. Talbot, and D. Cipollone. Geneva: United Nations Publishing.

Singh, Michael, Jane Kenway, and Michael W. Apple, eds. 2005. *Globalizing Education*. New York: Peter Lang.

Sipple, John, Kieran Kileen, and David H. Monk. 2004. "Adoption and Adaptation: School District Responses to State Imposed Learning and Graduation Requirements." *Educational Evaluation and Policy Analysis* 26 (2): 143–168.

Sirin, Selcuk, and Michelle Fine. 2008. *Muslim American-Youth: Understanding Hyphenated Identities through Multiple Methods*. New York: New York University Press.

Sirleaf, Ellen J. 2006. "Liberia's Gender-Based Violence National Action Plan." *Forced Migration Review* 27: 34.

Sizer, Theodore R. 1984. *Horace's Compromise: The Dilemma of the American High School*. Boston: Houghton Mifflin.

———. 1992. *Horace's School: Redesigning the American High School*. Boston: Houghton Mifflin.

———. 1996. *Horace's Hope: What Works for the American High School*. Boston: Houghton Mifflin.

Smith, Dorothy E. 2005. *Institutional Ethnography: A Sociology for People*. New York: AltaMira Press.

Smith, Mary L. 2004. *Political Spectacle and the Fate of American Schools*. New York: Routledge.

Sperry, Paul. 2004. "Pakistani Travelers under New Scrutiny." *World Net Daily*, June 28. http://www.worldnetdaily.com/news/article.asp?ARTICLE_ID=39188

Spindler, George D. 1959. *The Transmission of American Culture*. Cambridge, MA: Graduate School of Education, Harvard University.

———. ed. 1963. *Education and Culture: Anthropological Approaches*. New York: Holt, Rinehart and Winston.

———. 1974. *Education and Cultural Process: Towards an Anthropology of Education*. New York: Holt, Rinehart and Winston.

———. 2000a. "The Four Careers of George and Louise Spindler: 1948–2000." *Annual Review of Anthropology* 29: xv–xxxviii.

———. 2000b. *Fifty Years of Anthropology and Education, 1950–2000: A Spindler Anthology*. New York: Taylor and Francis.

Spindler, George, and Louise S. Spindler. 1987. *Toward an Interpretive Anthropology of Education at Home and Abroad.* Mahwah, NJ: Lawrence Erlbaum.

St. Pierre, Elizabeth A., and Wanda S. Pillow. 2000. *Working the Ruins: Feminist Poststructural Theory and Methods in Education.* New York: Routledge.

Stambach, Amy. 1994. "'Here in Africa, We Teach; Students Listen': Lessons about Culture from Tanzania." *Journal of Curriculum and Supervision* 9 (4): 368–385.

———. 2000. *Lessons from Mount Kilimanjaro: Schooling, Community and Gender in East Africa.* New York: Routledge.

Stanton-Salazar, Ricardo D. 1997. "A Social Capital Framework for Understanding the Socialization of Racial Minority Children and Youths." *Harvard Educational Review* 67 (1): 1–36.

Steiner-Khamsi, Gita, ed. 2004. *The Global Politics of Educational Borrowing and Lending.* New York: Teachers College.

Stenhouse, Lawrence. 1979. "Case Study in Comparative Education: Particularity and Generalization." *Comparative Education* 20 (2): 5–11.

Sue, Derald W., Jennifer Bucceri, Annie I. Lin, Kevin L. Nadal, and Gina C. Torino. 2007. "Racial Microaggressions and the Asian American Experience." *Cultural Diversity and Ethnic Minority Psychology* 13 (1): 72–81.

Suleiman, Yasir, ed. 1996. *Language and Identity in the Middle East and North Africa.* Richmond, Surrey: Curzon Press.

———. 2003. *The Arabic Language and National Identity: A Study in Ideology.* Washington, DC: Georgetown University Press.

———. 2004. *A War of Words: Language and Conflict in the Middle East.* Cambridge: Cambridge University Press.

Sunderman, Gail L. 2007. *Supplemental Educational Services Under NCLB: Charting Implementation.* Los Angeles, CA: The Civil Rights Project, UCLA.

Sutton, Margaret, and Bradley A. U. Levinson, eds. 2001. *Policy as Practice: Towards a Comparative Sociocultural Analysis of Educational Policy.* Westport, CT: Greenwood.

Swedish International Development Cooperation Agency (SIDA) 2003. *Investing in Future Generations.* http://www.sida.se/sida/jsp/sida.jsp?d=118&a=2806&language=en_US&searchWords=investing%20for%20future%20generations

Swenson, Carol R. 1998. "Clinical Social Work's Contribution to a Social Justice Perspective." *Social Work* 43: 527–537.

Tanzania Commission for AIDS [TACAIDS]. 2008. *UNGASS Country Progress Report: Tanzania Mainland.* Dar es Salaam: TACAIDS.

Taylor, James R., and Elizabeth J. Van Every, eds. 2000. *The Emergent Organization: Communication as Its Site and Surface.* Mahwah, NJ: Lawrence Erlbaum.

Thalhammer, Eva, Vlasta Zucha, Edith Enzenhofer, Brigitte Salfinger, and Günther Ogris. 2001. *Attitudes towards Minority Groups in the European Union: A Special Analysis of the Eurobarometer 2000 Survey on Behalf of the European Monitoring Centre on Racism and Xenophobia.* Vienna, Austria: SORA Vienna.

Tomlinson, Kathryn, and Pauline Benefield. 2005. *Education and Conflict: Research and Possibilities.* http://www.nfer.ac.uk/publications/pdfs/ ecoreport.pdf

Torgerson, Douglas. 2003. "Rethinking Policy Discourse." In *Deliberative Policy Analysis: Understanding Governance in the Network Society*, ed. M. A. Hajer and H. Wagenaar. Cambridge: Cambridge University Press.

Troyna, Barry. 1993. *Racism and Education: Research Perspectives*. Buckingham: Open University Press.

Tsing, Anna Lowenhaupt. 2005. *Friction: An Ethnography of Global Connection*. Princeton, NJ: Princeton University Press.

Tufte, Thomas. 2002. *Femina Health Information Project: September 1999–January 2002*. Stockholm, Sweden: Swedish International Development Corporation.

Tyack, David, and Larry Cuban. 1995. *Tinkering towards Utopia: A Century of Public School Reform*. Cambridge, MA: Harvard University Press.

Tyler, Tom R. 1997. *Social Justice in a Diverse Society*. Boulder, CO: Westview.

UNAIDS. 2008. *2008 Report on the Global AIDS Epidemic*. http://www.unaids.org/en/KnowledgeCentre/HIVData/GlobalReport/2008/2008_Global_report.asp

UNESCO. 2006. *HIV and AIDS Education: Teacher Training and Teaching—A Web-Based Desk Study of 10 African Countries*. http://www.popline.org/docs/1733/313126.html

United Nations Development Programme. 2006. *National Human Development Report Liberia: Mobilizing Capacity for Reconstruction and Development*. http://www.lr.undp.org/NHDR%2706_web.pdf

———. 2007. *Human Development Report 2007/2008*. http://hdr.undp.org/en/reports/global/hdr2007–2008/

United Republic of Tanzania. 1998. *Policy Paper on Local Government Reform*. Dar es Salaam: President's Office of Regional Administration and Local Government (PORALG).

———. 2001. *Education Sector Development Programme: Primary Education Development Plan (2002–2006)*. Dar es Salaam: Ministry of Education and Culture, Basic Education Development Committee (BEDC).

———. 2005. *Local Government Reform Program Progress Report January–June 2005*. Dar es Salaam: President's Office of Regional Administration and Local Government.

U.S. Department of Education. 2001. *No Child Left Behind Act of 2001*. http://www.ed.gov/nclb/landing.jhtml

Valdiviezo, Laura Alicia. 2006. "Interculturality for Afro-Peruvians: Towards a Racially Inclusive Education in Peru." *International Education Journal* 7 (1): 26–35.

———. 2009. "Bilingual Intercultural Education in Indigenous Schools: An Ethnography of Teacher Interpretations of Government Policy." *International Journal of Bilingual Education and Bilingualism* 12 (1): 61–79.

Valenzuela, Angela. 1999. *Subtractive Schooling: US-Mexican Youth and the Politics of Caring*. Albany: State University of New York Press.

Van Dijk, Teun A. 1997. "Political Discourse and Racism: Describing Others in Western Parliaments." In *The Language and Politics of Exclusion: Others in Discourse*, ed. S. H. Riggins. Thousand Oaks, CA: Sage.

Vavrus, Frances. 2003. *Desire and Decline: Schooling amid Crisis in Tanzania*. New York: Peter Lang.

————. 2005. "Adjusting Inequality: Education and Structural Adjustment Policies in Tanzania." *Harvard Educational Review* 75 (2): 174–201.

Vavrus, Frances, and Lesley Bartlett. 2006. "Comparatively Knowing: Making a Case for the Vertical Case Study." *Current Issues in Comparative Education* 8 (2): 95–103.

Vertovec, Steven. 1999. "Conceiving and Researching Transnationalism." *Ethnic and Racial Studies,* 22 (2): 447–462.

Vlopp, Leti. 2002. "The Citizen and the Terrorist." *UCLA Law Review,* 49: 1575–1600.

Vulliamy, Graham. 1990. "The Potential of Qualitative Educational Research in Developing Countries." In *Doing Educational Research in Developing Countries: Qualitative Strategies,* ed. G. Vulliamy, K. Lewin, and D. Stephens. London: Falmer Press.

————. 2004. "The Impact of Globalisation on Qualitative Research in Comparative and International Education." *Compare* 34 (3): 261–284.

Wampler, Brian. 2003. "Orçamento Participativo: uma Explicação para as Amplas Variações Nos Resultados." In *O Orçamento Participativo e a Teoria Democrática: O OP em Balanço Crítico,* ed. L. Avritzer. São Paulo: Cortez.

Ward, Tanya. 2004. *Education and the Language Needs of Separated Children.* Dublin: City of Dublin Vocational Educational Committee.

Watchlist. 2002. *Angola: Issue 2.* New York: Watchlist.

Werbner, Pnina. 2004. "Theorising Complex Diasporas: Purity and Hybridity in the South Asian Public Sphere in Britain." *Journal of Ethnic and Migration Studies* 30 (5): 895–911.

White, Sarah C. 1996. "Depolitising Development: The Uses and Abuses of Participation." *Development in Practice* 6 (1): 6–15.

Wildman, Terry M., and Jerry A. Niles. 1987. "Essentials of Professional Growth." *Educational Leadership* 44 (5): 4–10.

Williams, Raymond. 1983. *Keywords: A Vocabulary of Culture and Society.* Oxford: Oxford University Press.

Willis, Paul. 1981. *Learning to Labor: How Working Class Kids Get Working Class Jobs.* New York: Columbia University Press.

Winthrop, Rebecca, and Jackie Kirk. 2005. "Teacher Development and Student Well-Being." *Forced Migration Review* 22: 18–21.

Women's Commission for Refugee Women and Children. 2006. *Help Us Help Ourselves: Education in the Conflict to Post-Conflict Transition in Liberia.* http://www.womenscommission.org/pdf/lr_ed.pdf

Wood, George. 2004. "A View from the Field: NCLB's Effects on Classrooms and Schools." In *Many Children Left Behind: How the No Child Left Behind Act Is Damaging Our Children and Our Schools,* ed. D. Meier and G. Wood. Boston: Beacon.

World Bank. 1992. *Révitalisation de l'Enseignement Supérieur au Sénégal: Les Enjeux de la Réforme* [Revitalization of Higher Education in Senegal: The Stakes of the Reform]. Dakar: World Bank.

————. 1996. *Staff Appraisal Report: Republic of Senegal, Higher Education Project.* Washington, DC: World Bank.

World Bank. 2000. *Higher Education in Developing Countries: Peril and Promise.* Washington, DC: World Bank.

———. 2001. *Report and Recommendation of the President of the International Development Association to the Executive Directors on a Proposed Credit in the Amount of US$150 Million to the United Republic of Tanzania for a Primary Education Development Program.* Report No: P7466 TA. Washington, DC: World Bank.

———. 2003. *Implementation Completion Report on a Credit in the Amount of US$26.5 Million to the Republic of Senegal for a Higher Education Project.* Washington, DC: World Bank.

———. 2005. *Indigenous Peoples, Poverty and Human Development in Latin America: 1994–2004 (Executive Summary)* http://www-wds.worldbank.org/external/default/main?pagePK=64193027&piPK=64187937&theSitePK=5236 79&menuPK=64187510&searchMenuPK=64187283&theSitePK=523679& entityID=000112742_20050720172003&searchMenuPK=64187283&theSite PK=523679

———. 2008. *Liberia Education Country Status Report: Concept Note.* Unpublished report.

Yanow, Dvora. 2000. *Conducting Interpretive Policy Analysis.* London: Sage.

Yuval-Davis, Nira, Floya Anthias, and Eleonore Kofman. 2005. "Secure Borders and Safe Haven and Gendered Politics of Belonging: Beyond Social Cohesion." *Ethnic and Racial Studies* 28 (3): 513–535.

Zakharia, Zeena. 2009. "Language-in-Education Policies in Contemporary Lebanon: Youth Perspectives." In *Trajectories of Education in the Arab World: Legacies and Challenges,* ed. O. Abi-Mershed. New York and London: Routledge.

Zhou, Min, and Carl L. Bankston. 1998. *Growing up American: How Vietnamese Children Adapt to Life in the United States.* New York: Russell Sage Foundation.

Contributors

Lesley Bartlett is Associate Professor of Anthropology and Education at Teachers College, Columbia University. Her research and teaching interests include sociocultural studies of literacy, migration and education, and teacher education in international contexts. She is the author of *The Word and the World: The Cultural Politics of Literacy in Brazil*, co-author of *Additive Schooling in Subtractive Times: Bilingual Education and Dominican Youth in the Heights*, and co-editor (with Frances Vavrus) of *Teaching in Tension: International Pedagogies, National Policies, and Teachers' Practices in Tanzania*.

Audrey Bryan teaches International Educational Development, Development Education, and Comparative Education in the School of Education, University College Dublin. Her scholarly interests include globalization and education, multicultural and anti-racist education, and international educational development policy and practice.

Ameena Ghaffar-Kucher is Senior Lecturer in the Education, Culture, and Society division and Associate Director of the International Educational Development Program at the University of Pennsylvania. Her research centers on immigrants and education, trans/nationalism, citizenship, and curriculum and pedagogy in Muslim majority contexts, as well as the greater New York City area.

Maria Hantzopoulos is Assistant Professor of Education at Vassar College and a participating faculty member in the Programs in International Studies, Urban Studies, and Women's Studies. She also coordinates the Secondary Education Certification Program. Maria recently co-edited the book (with Alia Tyner-Mullings) *Critical Small Schools: Beyond Privatization in Urban Educational Reform*. She recently received the Social and Science Research Council and British Council's "Our Shared Past" grant to fund a larger research project on US World History textbooks, of which she is principal investigator.

Jill Koyama is Assistant Professor in the department of Educational Leadership and Policy, SUNY at Buffalo. She studies issues of education policy. Her book *Making Failure Pay: High-Stakes Testing, For-Profit Tutoring, and Public Schools* was published in 2010. Other work appears in the *Journal of Education Policy, Educational Policy, Anthropology and Education*, and *Educational Researcher*.

Rosemary Max is Director, Office for International Programs at Loyola University Chicago. Her research focuses on higher education policy and university reform in Africa.

Mary Mendenhall is the Director of the International Rescue Committee-University of Nairobi Partnership for Education in Emergencies. Mendenhall's professional and research interests examine the quality, relevance and sustainability of educational support provided by international organizations for displaced students in conflict-affected contexts.

As AFS-USA's Director of School Outreach and Educational Partnerships, **Tonya Muro** works to establish AFS-USA as a thought leader in global competency and 21st century educational skills. Tonya cultivates partnerships with like-minded organizations and develops intercultural learning resources for secondary school educators and their classrooms. Prior to joining AFS-USA, Tonya was the Program Director at Global Nomads Group, an NGO that fosters dialogue and understanding amongst the world's youth. Tonya holds an EdD and MA in international educational development from Columbia University's Teachers College, where she was a Fulbrighter in Tanzania.

Janet Shriberg is a former Assistant Professor in the International Disaster Psychology Program at the University of Denver. Her research and practice areas include increasing support for caregivers working in postwar settings, child protection, and preventing gender-based violence in situations of forced displacement. She has worked with local and international humanitarian organizations for more than a decade.

Aleesha Taylor is Deputy Director of the Open Society Foundations' Education Support Program. Prior to joining OSI in July 2007 as a Senior Program Manager, Aleesha was a Lecturer in the Department of International and Transcultural Studies at Teachers College, Columbia University, where she also completed her doctoral studies. Aleesha also holds degrees in psychology from Spelman College and the Graduate Faculty of the New School for Social Research, New York. She is a Term Member of the Council of Foreign Relations.

Laura Alicia Valdiviezo's ethnographic research focuses on bilingual and Indigenous education, sociocultural approaches to language policy design and implementation, and multicultural education in the Americas. Laura is Assistant Professor of Language, Literacy and Culture in the Teacher Education and Curriculum Studies Program at the University of Massachusetts, Amherst.

Frances Vavrus is McKnight Presidential Fellow and Associate Professor in the Department of Organizational Leadership, Policy, and Development at the University of Minnesota, where she also serves as Coordinator of the Program in Comparative and International Development Education. Her research focuses on development education and foreign aid policy, teacher education in sub-Saharan Africa, and gender studies. Her previous publications include her book *Desire and Decline: Schooling amid Crisis in Tanzania* and *Critical Approaches to Comparative Education* (co-edited with Lesley Bartlett) in addition to articles in *Comparative Education Review, Gender and Education*, and the *Harvard Educational Review*.

Moira Wilkinson is an independent education consultant. Her work focuses on educational quality, gender equality, and sustainability. Her current research examines the interrelationship among gender, decision-making and money. Find more information at moirawilkinson.com.

Zeena Zakharia is Assistant Professor of Comparative Education at the University of Massachusetts, Boston. She was the Middle Eastern Studies Postdoctoral Fellow at Columbia University and Tueni Fellow at the Carr Center for Human Rights Policy at Harvard's John F. Kennedy School of Government. Her publications examine the interplay of language, conflict, and peace-building in education. These interests stem from over a decade of educational leadership in war-affected contexts.

Index

Printed and bound by CPI Group (UK) Ltd, Croydon, CR0 4YY